CURRICULUM
.LINKS

ages
5–7

D1797311

Sound

Suzanne Kirk

Credits

Author
Suzanne Kirk

Editor
Christine Harvey

Assistant Editor
Charlotte Ronalds

Series designer
Lynne Joesbury

Designer
Catherine Mason

Illustrations
Gaynor Berry

Cover photographs
2004/Photo Network/Alamy.com

Photographic symbols
Music © Stockbyte.
Science © Ingram.

Published by Scholastic Ltd,
Villiers House,
Clarendon Avenue,
Leamington Spa,
Warwickshire
CV32 5PR
Printed by Bell & Bain Ltd, Glasgow
Text © Suzanne Kirk
© 2004 Scholastic Ltd
1 2 3 4 5 6 7 8 9 0   4 5 6 7 8 9 0 1 2 3

Visit our website at www.scholastic.co.uk

British Library Cataloguing-in-Publication Data
A catalogue record for this book is available from
the British Library.

ISBN 0-439-971241-1

Contents

Acknowledgements

Photographs
page 5 © Fotosonline/Alamy
page 9 © 2004/David Hoffman Photo Library/Alamy
page 20 © 2004/Sally and Richard Greenhill/Alamy
page 23 © Photodisc, Inc.
page 26 © SODA
page 29 © Ingram Publishing
page 34 © 2004/Photofusion Picture Library/Alamy
page 39 © 2004/David Stares/Alamy
page 41 © 2004/Photofusion Picture Library/Alamy
page 45 © Ingram Publishing
page 53 © Ingram Publishing
page 57 © Derek Cooknell
page 60 © 2004/Sally and Richard Greenhill/Alamy

Introduction

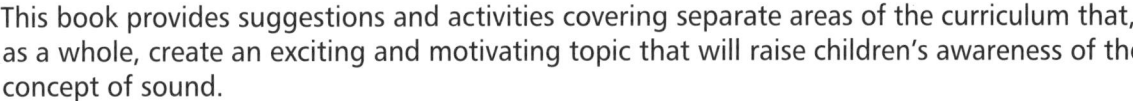

This book provides suggestions and activities covering separate areas of the curriculum that, as a whole, create an exciting and motivating topic that will raise children's awareness of the concept of sound.

Sound brings together aspects of science and music. It will help you to present an interesting and relevant topic at Key Stage 1 over a number of weeks, enabling children to explore and investigate aspects of sound.

Making sounds and music is an important part of everyday life. Throughout all ages and cultures, people have instinctively made sounds and music. They have collected natural objects and discovered different ways of making sounds for communication, as a warning and purely for enjoyment. Enquiring minds have also tried to solve the mystery of sound and to explain how we hear.

What subject areas are covered?
This book covers the QCA science Scheme of Work unit 1F 'Sound and hearing', and music unit 2 'Sounds interesting – Exploring sounds'. There are valuable opportunities for scientific observation and investigation as well as musical exploration. Generally, the activities in each section follow on progressively. In Section 5 the activities follow the sequence of a science investigation. Section 7 brings together the children's experiences and explorations in science and music, and enables them to illustrate a story or poem in an expressive way using sound effects.

Teaching specific subject areas through a topic
While it is important to distinguish between the separate subject areas of science and music, the natural links can be difficult to ignore and, in fact, are extremely useful in relating one area of work with another. One subject focus can provide an opportunity for exploration in another field.

There are many areas where science and music overlap. While children are enjoying exploring music, both with their voices and with instruments, their curiosity and interest in the sounds they produce can lead them to ask questions, which in turn can lead to a scientific experiment.

At this stage of their development children are increasing their understanding of sounds. They are becoming aware of the huge variety of sounds and sound sources that they encounter in their everyday experiences. They relate sounds to their sense of hearing and are introduced to the idea that sounds travel away from their source.

A carefully planned topic can meld together prescribed areas of the curriculum to create interesting learning opportunities appropriate to the needs of the class. Topic-based work presents a whole picture, motivates children and encourages their enthusiasm.

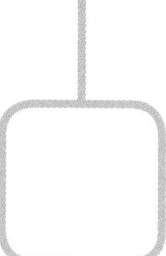

Getting started

Find out about any children in the class with hearing difficulties. Children with a serious hearing impairment will need visual support for many of the activities. Be aware of children with minor problems. There might be some struggling with temporary, or as yet undiagnosed, hearing difficulties. Be prepared to adapt activities as necessary to show sensitivity towards children with problems. Encourage all the children to behave in a sensitive manner towards each other. Many opportunities will arise for leading by example, demonstrating support, care and thoughtfulness towards others.

Locate any musical instruments within the school that the children can use. Try to obtain more unusual instruments for the children to observe, perhaps from different cultures and made from different materials.

Plan a code of conduct for the children when working with instruments. Decide when they can explore and experiment, and when they are restricted to observing the instruments on display. Be prepared to encourage focused listening when appropriate and devise signals, both audible and visual, when you need the children's attention.

Involving parents and carers

It is important that parents and carers are encouraged to take an interest in the topic their children will be focusing on. Involving parents and carers is a useful strategy in helping to motivate children. Prepare a letter for the children to take home, outlining the areas of work. Ask for extra information concerning hearing problems and be prepared to talk to parents and carers with concerns relating to school work.

Explain that the children will want to share their ideas and practise making sounds at home. Parents and carers can encourage children to listen carefully, to identify and recognise different sounds, and support their musical aspirations. They can listen to their children's questions and help with any research tasks.

If appropriate, take this opportunity to request any help required when the children are working outside or in small groups.

Promise to invite parents and carers to view the children's work at the end of the topic and perhaps to enjoy a musical performance by the children.

Introducing the topic to the children

Present this topic as an exciting and enjoyable area of work for both yourself and the children. Make sure they know it is something to look forward to with lots of interesting activities. Explain to the children that they will be exploring and discovering new things relating to both science and music. Tell them that sometimes they will be working outside of the classroom.

Tell the children that there will be opportunities for making sounds and music, including singing and playing musical instruments. There will be games to play and the chance to perform at the end of the topic. Explain that there will also be scientific tasks to take part in. These will involve recording careful measurements and observations so that discoveries can be made.

Prepare the children for any difficulties that might arise relating to children with a hearing impairment. Explain that some people cannot see as well as others and need glasses to help them. In a similar way, there are people who do not hear as well as others and they also need help in different ways. Encourage the children to talk about relatives, or people they know, who have problems with their hearing.

Starting points

Elicit the children's response to sounds. What do they think of when they hear the word 'sound'? Talk about some of the sounds you like to hear, such as children singing, people laughing, favourite music, and so on. Point out that you also like the opposite of sounds.

Ask the children when they think you like silence, or peace and quiet. Ask the children if they think they are good at listening. Do they always hear what is said to them?

Scientific observation and investigation

The content of science QCA unit 1F 'Sound and hearing' provides opportunities for the children to make observations and take part in a class investigation. As well as looking, observations include hearing and touching. In most cases observations are recorded. Recording science work can be in the form of drawings or diagrams, notes or sentences, or sometimes an account.

The activities in this book encourage the children to think and enquire. Asking questions is an important part of science investigation.

The activities in Section 5 set out ideas for a class investigation, designed to show that sounds get fainter as they travel away from a source. The children begin by taking part in an experiment from which questions develop. They then follow the stages of an investigation – planning together as a class, making observations, taking measurements, presenting results, making comparisons and drawing a conclusion. It is important to encourage a scientific method of working. Emphasise the importance of accurate measurements and careful observations, as well as the thoroughness required for the test to be fair.

When working outside, ensure the children behave appropriately. Children are understandably excited about leaving the classroom, but they should appreciate that they are still carrying out science work, in a different environment where they can explore, investigate and discover.

Music exploration

The emphasis of the music activities should be enjoyment. Try to repeat songs and sound games at different times during the week when there are spare moments. It might not always be necessary for the children to record themselves as part of an activity.

Be aware of any children showing exceptional musical ability and extend the activities accordingly.

Vocabulary

Both the science and music activities require specific, as well as frequently used, vocabulary. For instance, words that are commonly used throughout the activities to describe making sounds are *tap*, *bang*, *scrape*, *blow* and *shake*. It might be useful to list groups of linked words like these and make them available to the children as individual booklets, a class book or as a wall display.

Resources

Pictures showing:
■ different sound sources
■ a wide range of musical instruments
■ animals that make different sounds
■ scenes where different sounds could be heard, such as a seashore, a woodland, a railway station, swimming baths
■ people conducting choirs and orchestras
■ pictures or cartoons representing loud noises.

Books:
■ non-fiction books relating to sound and hearing, instruments, music in different cultures
■ stories with sound sequences, dramatic and noisy events or musical settings
■ a story or poem to act out as a stimulus for providing expressive sounds to accompany it.

Songs:
■ a bank of songs with which the children are familiar.

Getting started

Recordings:
- everyday and more unusual sounds
- a sequence of sounds from a specific location, such as a town, a seaside, a woodland
- a specially made recording of familiar voices
- a road safety video clip
- music to demonstrate different ways of making sounds, such as fast and slow, loud and quiet.

As well as standard classroom equipment, collect together the following:
- everyday items to make sounds with, such as a pencil or a spoon
- musical instruments, including some from different cultures
- recording equipment
- earmuffs or ear protectors
- small models of a fire engine or police car and pedestrians
- art materials
- measuring/metre sticks
- clipboards for outdoor work
- a camera and video camera.

The following are not essential, but would add value to the topic:
- a storyteller
- a musician/percussionist
- visits by people from a local theatre or radio station who can demonstrate sound effects.

Listening for sounds

FOCUS

SCIENCE
- recognising everyday sounds
- using the sense of hearing
- making observations of sounds, including voices

SCIENCE

MUSIC
- recognising different sound sources
- describing different sounds
- recognising different sounds in music

MUSIC

ACTIVITY 1

A LISTENING WALK

SCIENCE   MUSIC

Learning objective
To recognise different sound sources.

Resources
Three large sheets of paper; clipboard, paper and pencil for out of classroom recording; pictures of places where a different set of sound sources could be heard, such as a wood, the seashore, a swimming pool, a town; photocopiable page 17.

Preparation
You will need to provide three opportunities or areas for the children to listen, such as inside the classroom, another part of the school and an outside area. Write a heading on each of the three sheets of paper describing each listening area.

Activity
Talk to the children about listening and hearing. First ask them to touch the parts of their body they use for hearing. Depending on previous work, remind them that their ears are sense organs and that hearing is one of their five senses. Point out that they are using their ears to recognise sounds all the time, and at this very moment they are using them to recognise your voice.

Tell the children that they are going to concentrate on listening during this activity, so at certain times they will all need to remain as quiet as possible. Suggest that they might like to close their eyes to help them focus on listening and not be distracted by seeing things around them.

© 2004/David Hoffman Photo Library/Alamy

1

Listening
for sounds

Start in the classroom. Ask the children to listen very carefully and try to remember any sounds they hear during the next few minutes while they remain silent. Afterwards, use the prepared sheet to make a list of the sources from which the sounds they heard came. These might include voices from a neighbouring room, a ticking clock, a door or chairs moving, a cough, traffic passing by, footsteps. Ask the children to think of normal classroom sounds when everyone is busy. Suggestions might include the sound of voices, adults talking, things falling to the floor, a tap running, classroom equipment being moved. Point out that it is the children themselves who are the source of most of the sounds as they perform their classroom tasks.

Take the children to their next venue, which might be a different part of the school (you will need a clipboard and paper if so). Allow a few minutes listening time and then write down the sounds the children have heard. Emphasise the importance of recording observations to the children, explaining that it is easy to forget afterwards.

Next, move on to your third listening area, perhaps an outdoor area, such as the playground or a distant part of the school field. Encourage focused listening again, and then ask the children to recall the sounds they heard. Record their observations.

Back in the classroom, transfer the observations you recorded onto the large sheets of paper. Encourage the children to make comparisons between the sounds listed on the sheets. Were there any sounds common to all three areas? Which sounds were heard only outside? Then ask the children to remember the sounds they heard and to describe them. Which was the loudest sound? The quietest? Was there a continuous sound? Was there a silent time when the children did not hear any sounds at all? Which of the listening areas was the quietest? Is it easy to decide? Would the children hear the same sounds in each of the places if they listened at different times? Do they think there is an especially noisy time anywhere, such as at rush hour? What would listening at night be like?

Ask the children to think about sounds they would hear in different places, such as a park, a swimming pool, the middle of a town, the seaside, the woods. You could use pictures to act as stimuli if appropriate.

Recording
Provide each child with a sheet of paper divided into three parts, with headings relating to the three areas where the focused listening took place. Ask the children to choose several sound sources from each venue to draw and label. Some children could write sentences to describe the sounds in each area, making simple comparisons. Alternatively, provide the children with photocopiable page 17 which shows a garden scene. Ask the children to write the names of the sound sources they recognise in the picture and decide whether they think this garden is a noisy or a quiet place.

Differentiation
Children:
■ listen carefully and recognise different sound sources, recording with drawings and labels
■ listen carefully, recognising a range of sound sources and recalling sounds heard in different places; record with drawings and labels, making simple comparisons
■ focus their listening, recognising a range of sound sources; recall sounds heard in different venues; recording with drawings and labels, making relevant comparisons.

Plenary
Point out to the children that sounds are everywhere and that we hear different sounds in different places. Comment on quiet times. Where would the children go to enjoy peace and quiet? Suggest they carry out a listening experiment at home. They could listen carefully

inside their house, perhaps in the kitchen, and then outside, perhaps on their way to school. You could provide a sheet of paper for the children to record indoor sounds on one side and outdoor sounds on the reverse. Some children might enjoy making their own tape recordings of sounds they hear at home or when on a day out. Suggest they bring their recordings along for the class to listen to.

Display
Begin to build up a montage of illustrations representing sound sources. Arrange the three large sheets of paper that record the sounds heard during the activity and add the children's drawings and other illustrations appropriately.

SCIENCE   MUSIC

ACTIVITY 2

WHAT IS MAKING THIS SOUND?

Learning objective
To know that there are many different sources of sounds.

Resources
A specially prepared recording of sounds for the children to identify. If possible, obtain a recording of sounds specific to one area, such as the seaside, a busy town, a woodland or a farmyard.

Preparation
Select the sounds on the recording so that the children can hear single, easily recognisable sounds to begin with, such as a bell ringing, birds twittering, a dog barking, children singing, a toy with a distinctive sound. Then move on to sounds that will be less obvious and perhaps more complex, like wind in the trees, water rushing, traffic noises.

Activity
Tell the children that they are about to hear some sounds, many of which they will recognise easily and some which they might have to think about. Play a single familiar sound and ask the children to explain what they think made the sound. Are they all in agreement? Perhaps this is a sound they hear every day and know instantly what is making it. Play other simple sounds for the children to identify. If appropriate, ask the children to spot things in the room which might have been used to make the sound. Ask questions, too, to stimulate ideas. For example, How do we know that the bird was not a cuckoo or a duck? Do you know any birds that twitter like this sound? Was the bell in the classroom used to make the sound you heard? (It was probably a bigger bell used to make the sound they heard on the recording, such as one in a church.)

Move on to less obvious sounds. Discuss with the children what might have made these noises. For instance, do they think the sound of water is a stream, rain or water rushing down a pipe? Perhaps they have heard similar sounds and can make comparisons. It is not always necessary to reach a final decision. Conclusions such as, *Tom thinks it could be a river rushing over rocks; Lee thinks it is more likely to be a drainpipe overflowing* are fine.

Recording
Ask the children to draw and label some of the sources of sound that they have heard. Perhaps ask them to listen to a different set of sounds and record as many as they can remember. If they listen to a set of sounds from a specific area they can build up a picture of the scene.

Listening for sounds

Differentiation
Children:
■ recognise the sources of familiar sounds, recording these with drawings and labels
■ recognise a range of sound sources, recording these with drawings and accurate labels
■ recognise a range of less familiar and more complex sound sources, recording these as drawings with detailed labelling.

Plenary
Point out to the children that we use our sense of hearing to recognise sounds even when we cannot see what is making the sounds. Explain that we know the sound of a bell or a special toy just by hearing them, or that we know it is raining without seeing the raindrops. We can also tell when there are people nearby, as we can hear the sounds their movements make.

Display
Arrange the children's work in a temporary way so it can be admired and discussed. Continue to develop the montage of pictures of sound sources.

SCIENCE   MUSIC

ACTIVITY 3

LISTENING TO VOICES

Learning objective
To make observations of sounds by listening carefully.

Resources
Recordings of familiar voices.

Preparation
Make a recording of children's voices and members of staff speaking. It might be possible to record the voice of each class member, otherwise use a group of children that the class will know and record their voices. Record some examples of the children speaking normally so that everyone will be able to recognise them from their voice. Then perhaps introduce some surprises to challenge the children, such as the voice of a member of staff or a baby's shout or gurgle. Suggest some children attempt to disguise their voices by speaking very slowly or quickly. Include any examples of a distinctive laugh. Record a cough and a person humming or singing. Make vocabulary cards with the names of voice sounds on, such as *speak*, *laugh*, *sing*, *hum*, *cough*, and so on.

Activity
Tell the children that you have a sound quiz for them to take part in. Ask them to listen carefully while you play the recording of the first voice. Elicit the children's comments. Are they surprised that the sound is a voice? Do they recognise whose voice it is? Play the recording of other voices the children know well. Can they

identify whose voices they are hearing? Do they find this an easy task? Perhaps some children have a different regional accent and are very easy to identify. Point out that even those speaking with the same accent can be recognised because the children know each other well and have learned to recognise each others' voices as well as their faces. They do not always have to see each other to know who is speaking.

Continue to play the sound recordings, asking the children questions. Did they recognise the adult among the children's voices easily? How did they know which was the voice of a baby?

Play the disguised voices. Are the children/adults still recognisable? Is anyone able to remain anonymous? Play the laugh, cough and humming sounds. Can the children recognise who is making these sounds? If someone hums or coughs is it more difficult to recognise them? Try this out by holding up the vocabulary cards and asking some children to make the sound with their voice. Did they sound like they usually sound? Would the other children recognise them from that sound if they couldn't see them?

Recording

Ask the children to make a vocabulary list of all the sounds they heard during the activity, such as speaking, laughing, coughing and humming. They can write sentences to describe the voices of friends and family. For example, 'My little sister has a very squeaky voice' or 'When he sings my grandad has a very deep voice'.

Differentiation

Children:
■ are able to listen carefully and recognise some familiar voices
■ are able to listen carefully and identify the voices of a range of different speakers
■ are able to listen carefully and accurately, identifying and describing the voices of a wide range of different speakers.

Plenary

Point out that we all have a distinctive voice which helps us to be recognised by each other. However, we can alter our normal voice if we choose.

Display

Arrange the vocabulary cards in a prominent position.

ACTIVITY 4

DESCRIBING SOUNDS

MUSIC

Learning objective

To describe sounds, some of which are heard and some remembered.

Resources

A collection of items that will make contrasting sounds, such as loud and soft, high and low, continuous and intermittent (a tin with marbles and one with plastic counters, bells, a whistle, a rattle, toys that make sounds); a board and writing materials; pencils, pens and crayons; prepared recording page; prepared vocabulary cards.

Preparation

On an A4 piece of paper, prepare a recording framework that has six boxes on it with the headings *loud, quiet, high, low, all the time* and *now and then*. Make several sets of vocabulary cards of these words too.

Listening for sounds

Activity

Tell the children you will make some sounds that you want them to describe. First of all you will use your voice in two different ways. Ask the children to listen carefully, not to make any sounds themselves but to put up their hands when they notice any difference. Then speak a few sentences, or read a short poem, perhaps two lines in a fairly loud voice and two lines speaking softly. Encourage the children to talk about the loud voice and the quiet or soft voice you used. Then show the children the tin containing a few marbles. Ask what

the sound will be like when the tin is shaken. The children should suggest it will be noisy, making quite a loud sound. Replace the marbles with plastic counters and ask the children to describe the sound that they will hear this time. They should expect it to be quieter. Choose a child to demonstrate the difference. Ask the children to think of some other loud and quiet sounds they know. Make a list of about five of each type on the board.

Tell the children to listen again. Make a high and a low sound, perhaps using two different bells or toys with contrasting sounds. Ask the children to describe the difference between the two sounds this time. Emphasise that one sound is high while the other is low. Write down a few examples of high and low sounds which the children suggest. Ask some children to speak in high voices and others to speak in low voices to emphasise the difference.

Encourage the children to listen again. Make a continuous sound, perhaps from an electrical toy, and then an intermittent sound, which could be a clock ticking or a pencil tapping. Ask the children to describe these sounds to each other and encourage them to make comparisons. Talk about some sounds being there all the time once they have begun, while others are heard now and then. Write down some examples of each of these types of sounds that the children remember from experience.

Finish with a game. Ask individuals to choose a vocabulary card and to make the type of sound it describes. They could use their voices or choose something suitable in the room. Can the other children guess which word was on the card? Alternatively, distribute the cards among the children. Suggest a sound source, or show a picture, and ask the children to hold up their card if it describes the sound made. For example, if you showed a picture of an aeroplane a child might hold up a card with *loud* or *all the time* on it. Similarly, giving 'whistle' as a sound source might prompt cards with *high* or *now and then* on them.

Recording

Show the children the recording framework. Explain that they should read the headings and draw and label some sources which produce each type of sound.

Differentiation

Children:

■ recognise and describe some differences between sounds, recording examples with drawings

■ describe a range of sound differences, both when listening and remembering, recording examples with drawings and headings

■ accurately describe and compare a range of sound differences, both when listening and remembering, recording personal examples as labelled drawings.

Plenary

Ask some children to suggest different ways for you and others to use their voices when speaking. For example, speaking in a high or low voice, a soft or loud voice. Demonstrate an intermittent sound by speaking with pauses between words and a continuous sound without any pauses.

Display

If space allows, draw and cut out six boxes and label them with the headings from the children's sheets. Include examples of sources which demonstrate the sound differences.

MUSIC

ACTIVITY 5

SOUNDS OF MUSIC

Learning objective

To recognise different sounds in music.

Resources

A piece, or pieces, of music demonstrating different aspects of sound, such as loud and quiet, high and low, fast and slow, continuous and intermittent sections; music playing equipment or recording equipment; paper; crayons or painting materials.

Preparation

Select music to demonstrate loud, quiet, high, low, fast, slow, continuous and intermittent aspects of sound. You could use a piano, keyboard or other instrument to play and record familiar tunes yourself. Alternatively, find recording(s) of suitable music.

Activity

Tell the children that you want them to focus their listening on a short piece of music. Play the music once without any comment or interruption. Then tell the children you will play the music again and this time ask them to listen out for differences. Suggest that they might recognise loud and quiet sections of the music, or high or low parts. Ask them to listen carefully to identify differences. Then encourage the children to describe what they have heard. Perhaps they have noticed that the music starts quietly but becomes louder. They might recognise some very low sounds in the middle of the piece.

Play the music, or part of the piece, again and ask the children to indicate with their hands when they hear the fast and slow parts. Suggest they flutter their fingers, or make circular movements with their hands, either slowly or faster depending on the speed of the music. Then, to demonstrate high and low parts of the music the children can make the same movement or wriggle their fingers above their heads or low down by their knees. Ask them how they might signal loud and quiet pieces and also music which is playing all the time and now and then.

Provide opportunities for groups of children to demonstrate to the rest of the class. Praise accurate listening and sensible behaviour.

Recording

Ask the children to consider all the sounds they have experienced during this and previous activities and to write an account of their favourite sounds. Alternatively, the children could use paints or crayons to represent the sounds they have listened to in the piece of music. Discuss how they might show, using patterns and shapes, loud or soft, high or low and fast or slow sounds.

Listening for sounds

Differentiation
Children:
■ recognise and describe some differences when they hear a piece of music
■ recognise and describe a range of differences when listening to pieces of music
■ begin to show musical ability when identifying and describing different sounds when listening to music.

Plenary
Point out that people who write music and play instruments use sound in different ways to make the music special and more interesting for the people who are listening. Suggest the children listen for differences when they hear music at home on the television, or when listening to their own music.

Display
Put up the children's accounts with the heading *Favourite sounds*. Add any artwork to accompany that in the boxes from the previous activity.

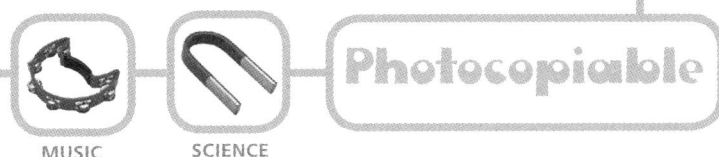

MUSIC   SCIENCE

Sounds all around

What sounds would you hear in this garden?

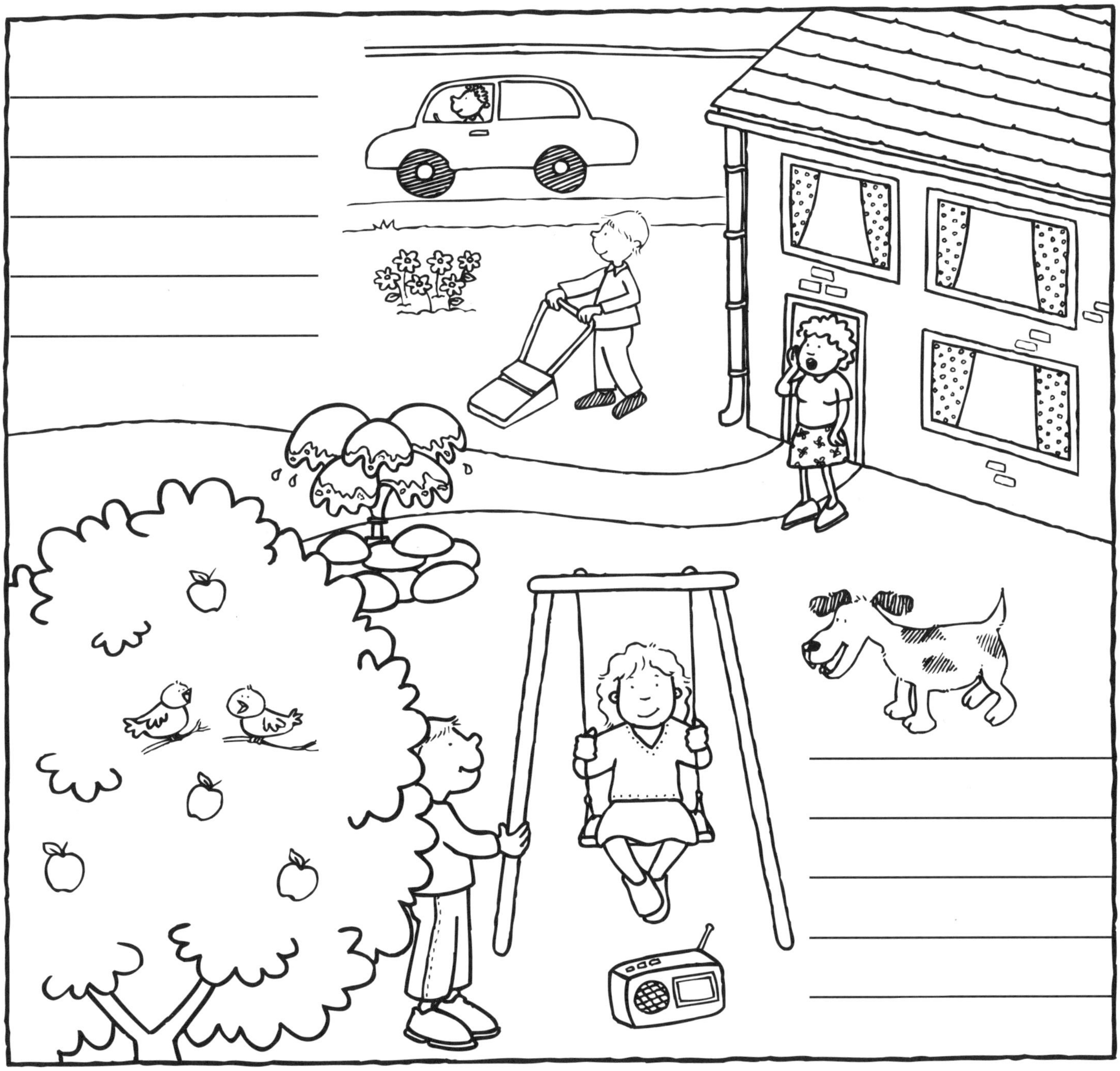

Is this garden a noisy or a quiet place?

■ S C H O L A S T I C

Section 2

FOCUS

SCIENCE
- making body sounds
- exploring ways of using voices
- understanding how sounds are made through touch and observation

MUSIC
- exploring sounds made using the body
- exploring sounds made using the voice

SCIENCE   MUSIC

ACTIVITY 1

BODY SOUNDS

Learning objective
To make sounds in different ways using parts of the body.

Resources
A board and writing materials.

Preparation
Decide on the organisation of the activity – how the children will know when to listen and when to make sounds. Consider the space available for when the children are practising sounds in pairs.

Activity
Begin by asking the children to listen carefully while you clap softly, tapping the fingers of one hand on the palm of your other. Ask the children what sort of a sound you have made and which parts of your body have been used. Identify the sound as a soft clapping, made by patting the palm of your hand gently with your fingers. Encourage the children to copy this sound accurately, perhaps changing from one hand to the other with three claps to each side.

Tell the children that you want them to explore different sounds they can make using different parts of their bodies. Ask for some ideas. These might include tapping different parts of their bodies, making sounds with their lips, teeth and tongue, or perhaps emphasising breathing sounds. Ask some children to demonstrate examples while the others watch and listen carefully. Then encourage the rest of the class to copy the sounds. Together decide what some of the sounds could represent. A popping sound with the lips could be water dripping down a pipe, a clicking sound

made with the tongue and cheeks could be a horse walking by.

If appropriate, give the children the opportunity to explore making further sounds in pairs. Set out a code of conduct so that the children work sensibly and without disturbing others. Suggest they discover and together practise two sounds made with different parts of their bodies, to show and describe to the rest of the class.

Bring the children together again and ask the pairs to demonstrate further examples of body sounds. Encourage the other children to describe their sounds. Make a list of the types of sounds that have developed, such as clap, tap, slap, pat, click, sniff, pop, squeak and stamp. Identify the parts of the bodies which have made the sounds and discuss how effective the sounds are. Would the children think that any examples were the real sound?

Recording
Ask the children to draw themselves making two (or more) different sounds. They should label the parts of the body they used and describe in words, as accurately as they can, the sounds they made.

Differentiation
Children:
■ explore sounds using different parts of their bodies
■ recognise how different parts of their bodies can be used to make sounds, describing sounds as words
■ explore sounds imaginatively using different parts of the body, describing sounds accurately in words.

Plenary
Recall the range of sounds made and the effects they produced. Talk about any particularly interesting examples. Point out that bodies can also be very quiet and movements can be silent. As the children move on to their next activity, ask them to proceed as quietly as possible.

Display
Put up a copy of the list of sounds the children made. At various times during the week, point to words at random and ask the children to make the sound.

ACTIVITY 2

VOICE SOUNDS

MUSIC

Learning objective
To explore different sounds that can be made by voices.

Resources
A board and different coloured writing materials; prepared sound sequences written out; prepared sheets for the children; A4 paper; coloured pencils; A4 recording sheets for the children, marked out with horizontal lines 1cm apart.

Preparation
Write on the board the sequence of sounds you intend to demonstrate at the beginning of the activity, using different colours to distinguish between short and long sounds. For example, *bbbb* in blue for short sounds and *oooo* in red for a long sound. Then cover until required using a drape or piece of paper with Blu-Tack.

2

Making
sounds

Activity

With the children in focused listening mode, start by repeating sounds with your voice. For example, *bbbbbbb*, *chchchchchchch*, *eeeeeee* or *oooooo*. Then show the children how you have described these sounds as letters on the board. Can the children work out why you have used different colours? Encourage the children to explore the sounds by repeating them with you to help them identify the long and short sounds.

Explain that our voices can make many different sounds. Ask individuals to demonstrate more sounds and encourage other children to suggest how they can be written. Add these to the list on the board.

Put the children into pairs and ask them to take turns to investigate other sounds and to write them down using different colours for long and short sounds. After about five minutes, bring the class back together. Ask the children to demonstrate some of their ideas. Develop the list of sounds on the board.

Next, ask the children to think of ways these sounds might be changed to make them more expressive. Demonstrate by making an *oooo* sound, first quietly then getting louder and then quieter again. Ask a child to demonstrate with another sound, making it louder and quieter. Give the rest of the class the chance to copy the sounds in the same way. Talk about the effectiveness of the changes.

Ask how else the sounds might be changed. Try faster and slower examples. Which sounds do the children enjoy making? What do the different sounds make them think of? For example, *This voice sound makes me think of a snake hissing* or *This voice sound reminds me of a car passing by*.

Suggest that sounds made with the children's voices could be mixed with sounds made by other parts of their bodies (as in the previous activity). Provide ideas for the children to try, such as *Make a long* aaaaa *sound with your voice. Now make some short sounds with your lips. Make some loud and then quieter sounds with your hands.*

The children will also enjoy making some of their sounds in a singing voice.

Divide the class into three groups and give each group a sound to make, perhaps a *shsh* voice sound, a tapping and a rubbing sound. Get the groups to perform their sounds at the same time. Rotate the sounds between the groups so that each group tries each sound. Describe the effect to the children. For example, *Together you sounded like a rainstorm*.

Recording

Provide the children with the marked out recording sheets. Ask them to use coloured pencils to make rows of their favourite sounds. Later, they can exchange pages and try out each other's sounds.

Differentiation

Children:
■ explore sounds using voices and recognise how sounds can be used expressively
■ explore sounds using voices, finding ways of changing sounds for different effects
■ explore sounds using voices, making changes imaginatively to create a range of effects.

Plenary

Ask the children to tell each other how they can change some of the sounds they can make with their voices and different parts of their bodies to create different effects. Ask some individuals to demonstrate their favourites.

Display

Use strings of letters representing sounds as borders to surround the children's work.

SCIENCE   MUSIC

ACTIVITY 3

USING VOICES

Learning objective
To explore ways of using voices.

Resources
A board and writing materials; photocopiable page 24; pencils and pens; a camera.

Preparation
Choose some songs the children know well and which can be sung in different ways.

Activity
Tell the children you want to make a list of all the different ways they can make sounds using their mouths. Start the list off with speaking. They should suggest singing, humming, whistling, coughing, and so on. Talk about each type of sound on the list. Discuss how singing is similar to talking, as words are often used. How do the children know when someone is singing rather than speaking? Can anyone demonstrate singing without making sense? For example, *dum-de-dum-de-dum* or *la la la*. How is humming different? Do the children think it is strange that they can hum with their mouths closed? What comments can the children make about whistling? Is it different from the other sounds? Is it their voice they are using or something else? Elicit comments about how the sound of a cough is made.

  Explain that our voices can be used to give different messages to other people. We can speak in different voices depending on how we feel and how we want others to feel. Discuss how we speak loudly or shout when we are angry, excited or trying to attract someone's attention. Discuss how we speak softly or perhaps whisper when we do not want everyone to hear, or when someone is asleep, upset or does not want to be disturbed. Talk about how we can sing in many different ways to communicate feelings and meanings to others.

  Encourage the children to sing a familiar song, at first in their usual way. Then suggest the song is sung again, but this time slowly, then quickly. Ask the children to sing again, this time loudly and then softly. Which type of singing do they think is best for this particular song? Which types of song should be sung quietly or softly? Suggest a lullaby sung to a sleepy baby. Do the children know a song which needs to be sung quickly, slowly or loudly? Are there other ways they could try singing to give people a different message or feeling?

  Divide the class into groups. Take close-up photographs of the faces of one group singing, one group speaking, one humming and one whistling, to use in a display later.

Recording
Give the children photocopiable page 24. Ask them to use words or sentences to describe their personal experiences of using their voices in different ways. Point out that the vocabulary list at the bottom of the page will help them to spell words correctly. Suggest they might like to draw small faces to show the facial expressions they make when creating each of these sounds.

Differentiation
Children:
■ recognise different sounds they can make with their voices and use their voices expressively
■ recognise different ways of using their voices and understand the significance of using their voices expressively
■ know the different ways in which they can use their voices and understand that using their voices expressively is an important part of communicating.

2

Making
sounds

Plenary

Reiterate that the sound our voices make when we are talking and singing can be changed to send different messages to others, perhaps to show how we feel.

Display

Use the photographs you took during the activity, or large pictures or drawings of faces, to show the expressions on faces when voices are used in different ways. Print out appropriate labels to accompany the faces.

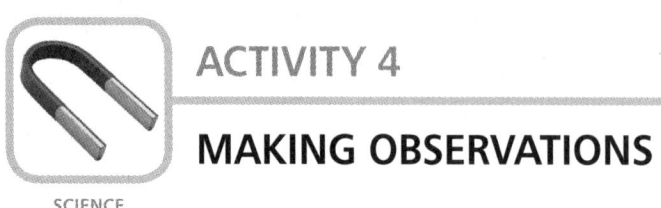

ACTIVITY 4

MAKING OBSERVATIONS

SCIENCE

Learning objective

To explore sounds by observation and touch

Resources

Small mirrors; photocopiable page 25; pictures of animals that make different sounds.

Activity

If possible, arrange the children in two rows facing each other. Tell them that they will be finding out what happens to their faces and throats when they make sounds. In turn, ask one row of children to say something. This could be a greeting, such as *Good morning, friends.* Ask the others to watch carefully to see what happens to the speakers' faces and throats. Ask the children to describe what they have discovered during their observations. Encourage them to talk about the movement of the lips and cheeks. Did the children see glimpses of teeth or tongues? What happened in the throat area?

Ask the rows of children, in turn, to sing a snatch of a song they know and then to describe their observations. Did they notice anything different or were the movements similar to when the children were speaking? Next, ask the children to whistle. What happens to their faces? Did the children see any movement of the throat? Finally, ask the children to hum. Did they notice any movement at all this time? Do they think it would be easy to identify one person humming in a group?

Now tell the children you want them to use their sense of touch to find out what happens when they make these sounds. This will involve feeling their own faces and throats with their fingers and must be done very gently.

Demonstrate how the children can use their fingertips to feel their faces and throats. While they remain silent, ask them to pat their cheeks. Talk about the softness of this part of the face. Move on, asking them to feel the bones of their jaws as they open and close their mouths and to investigate the hard and soft areas of their throats.

Ask the children to explore the movements of their jaw and throat while they speak. Encourage comments. What sort of movements can they feel? Which area has the most movement? Are there big movements or small wobbles? Which parts of the face and throat can be stretched? Then ask the children to sing, again while gently feeling the movements of their faces and throats. Do they notice any difference between the movements made when speaking and singing? Ask them to try humming. They should be able to tell you that there are no obvious movements as they hum, but they might feel a buzzing sensation (vibrations) if they hold their hand flat under their chins.

Ask the children to whistle while feeling their faces and throats. Perhaps they will be able to tell you that the sound made does not come from the throat but is made on the lips.

Recording
Provide the children with mirrors and ask them to describe the movements they see as well as what they feel when they are speaking, singing, whistling and humming. Using photocopiable page 25 they can record accurately what they have observed in writing.

Differentiation
Children:
■ are aware that parts of their bodies move when they speak, making observations and recording with words
■ recognise which parts of their bodies move when they speak, making observations by looking and feeling and recording with notes
■ recognise which parts of their bodies move when they speak, making and recording observations in a scientific manner by looking and feeling.

Plenary
Tell the children that they have made scientific observations by looking carefully and feeling their faces and throats to discover what happens when they make sounds with their voices. The movements their throats, mouths and faces make are essential so that we can speak, sing, whistle and hum. Point out that other animals make very different sounds from humans. Suggest the children might like to do some research at home. They can find out the names of the sounds different animals make. This information can be used to make a game in which animal and sound are matched.

Display
Use pictures of animals to create a wall puzzle. Print out the names of animal sounds for the children to use in a matching game.

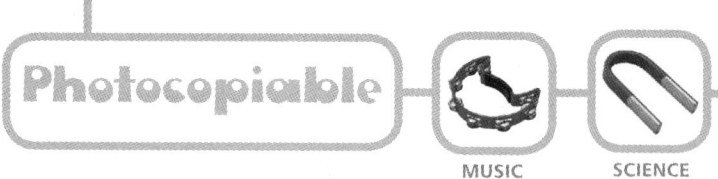

Using voices

Speaking

_____
_____
_____
_____

Whispering

_____
_____
_____
_____

Shouting

_____
_____
_____
_____

Whistling

_____
_____
_____
_____

Humming

_____
_____
_____
_____

Singing

_____
_____
_____
_____

loudly   slowly   sound   open   excited

favourite   song   mouth   quietly

angry

quickly

noise   sweetly   tune   closed

SCIENCE

Making observations

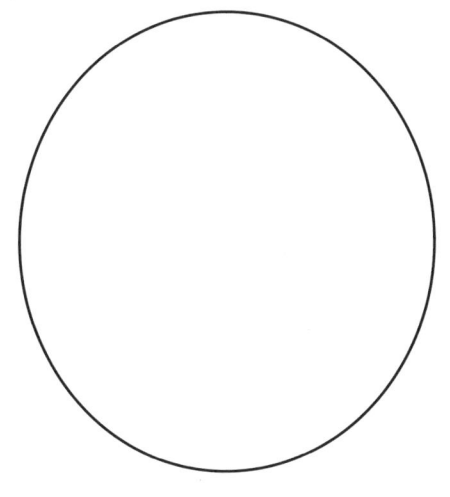

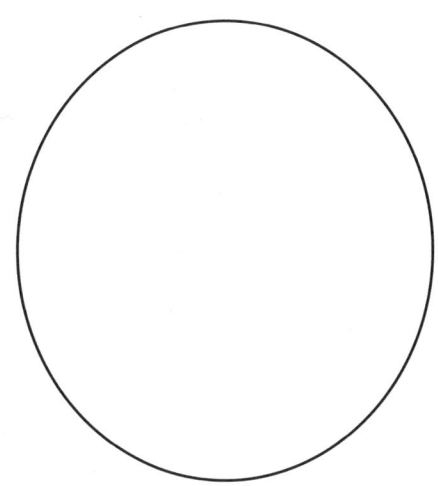

_____

_____

_____

_____

_____

_____

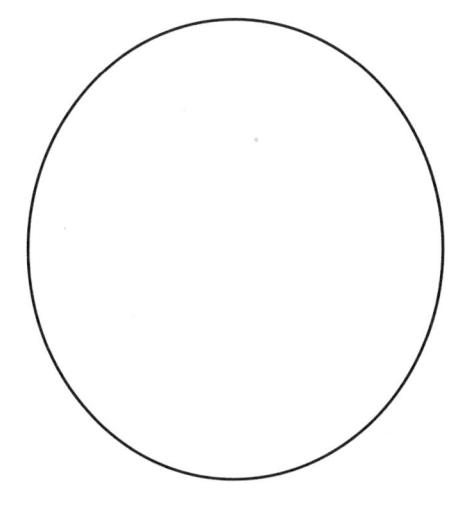

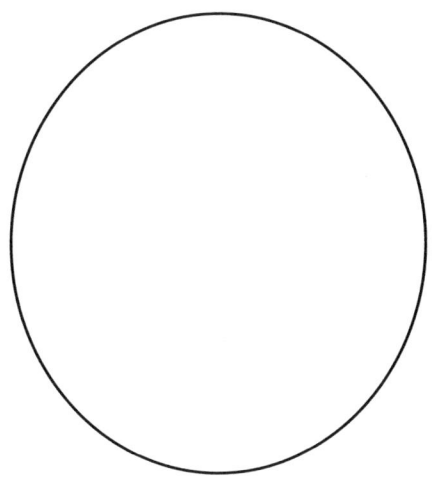

_____

_____

_____

_____

_____

_____

Enjoying sounds

FOCUS

MUSIC

MUSIC
- exploring ways of making sounds
- identifying and exploring simple instruments
- enjoying sounds
- recognising pleasant and unpleasant sounds

SCIENCE

SCIENCE
- using everyday objects to make sounds
- exploring instruments to make sounds
- presenting and interpreting results

SCIENCE   MUSIC

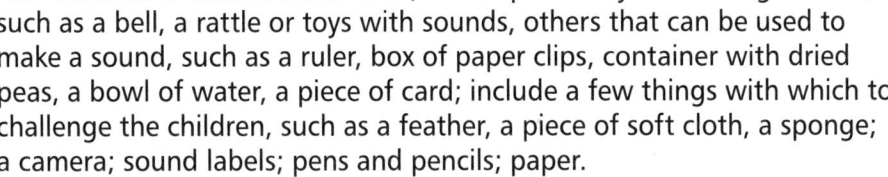

ACTIVITY 1

INTERESTING SOUNDS

Learning objective
There are many different ways of making sounds.

Resources
A range of items for the children to make sounds with, some specifically for making a noise, such as a bell, a rattle or toys with sounds, others that can be used to make a sound, such as a ruler, box of paper clips, container with dried peas, a bowl of water, a piece of card; include a few things with which to challenge the children, such as a feather, a piece of soft cloth, a sponge; a camera; sound labels; pens and pencils; paper.

Preparation
Plan the organisation of this activity, a part of which can take place in the school grounds. Also plan to take photographs of the children making their sounds. Make labels to describe ways of making sounds, including *tapping*, *ringing*, *rattling*, *scratching* and *buzzing*.

Activity
Tell the children that you want to find out what sounds can be made using some items you have collected. In turn, present individual children with an item and ask them to create a sound. For instance, give a child a rattle or a ruler and ask them to make a sound with it. Encourage the children to comment on the sounds made. Is the sound what they expected? Is it a pleasant sound? What words would they use to describe it? Perhaps it is a tapping, buzzing or swishing sound. The children might talk about the sound being loud, quiet, high or low. Ask the children if any item can make more than one sound. Are there any similar sounds? Did they think there would be so many different sounds that they could

© SODA

make with this collection of items? Can anyone produce a sound using a feather or a piece of sponge?

In pairs, ask the children to find an unusual way of making a sound using things around them. This could be done in the classroom or in a designated area of the school grounds. Set out any necessary ground rules and give the children ten to 15 minutes for their investigations.

When the children are together again, invite the pairs to demonstrate their sound. Encourage focused listening and ask for comments. Which are the most unusual items used to make a sound? Which is the most interesting sound? Which sounds are similar but made by different things? Do any of the sounds have anything in common? Perhaps most are made by tapping, some by shaking, a few by blowing. If appropriate, take photographs showing the children demonstrating their sounds.

At the end of the activity, put out the labels describing how sounds can be made and ask the children to place their items in the group they think is most appropriate. Ask the children to make simple drawings if the items used cannot be moved.

Recording
Ask the children to write an account of how they discovered different ways of making sounds in the classroom or out of doors. They can also write a comparison of two sounds, describing how each was made and the type of sound that was produced.

Differentiation
Children:
■ explore different ways of making sounds using everyday objects
■ explore different ways of making sounds using everyday objects, describing the sounds and making comparisons
■ explore different ways of making sounds using everyday objects, accurately describing the sounds and making comparisons.

Plenary
Point out that the children have discovered that there are many different ways of making interesting sounds. Most things can be used to make some kind of sound, which the children have discovered during their investigations.

Display
Match any photographs you took during the activity to the children's work. Encourage the children to add to the groups of items under the sound labels. Perhaps highlight a particular sound each week.

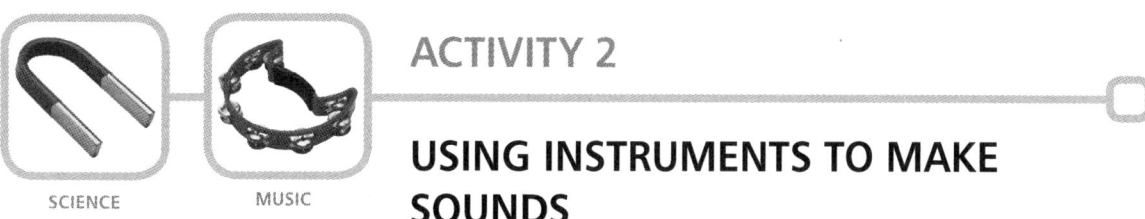

SCIENCE   MUSIC

ACTIVITY 2

USING INSTRUMENTS TO MAKE SOUNDS

Learning objective
To explore sounds made by musical instruments.

Resources
A selection of musical instruments, if possible one instrument for each child, to include percussion, recorders, whistle, keyboard, guitar or other instrument that is plucked,

3

instruments from different cultures; antibacterial wipes; pictures of instruments not available to the children; board and writing materials; photocopiable page 33; pens and pencils.

Preparation
Collect together and arrange the instruments for the children to observe and use. Plan how the procedure for the children to try out different instruments will work. Perhaps arrange the instruments in a circle with the children moving around to try each of them. Alternatively, make several small circles of instruments for groups of children to use. This part of the activity could take place out of doors.

Activity
Arrange a selection of musical instruments where the children can see them. Ask individuals, or pairs of children, to choose instruments and use them to make a sound. With each sound, ask the rest of the class how the sound was made. What action was required? Encourage the children to consider whether the instrument was banged or tapped. Perhaps it was shaken or strings were plucked. Which instruments needed blowing? Is there an instrument that can make a sound in more than one way?

Produce a new instrument and ask how the children think it might be played to make a sound. Try this out to see if they were right. With the children's help, write a list of all the actions needed to make sounds with the instruments. For example, tap, bang, scrape, blow, shake, beat, pluck.

If possible, provide the children with the opportunity to try out different instruments. Have the instruments arranged so that the children can try each one, moving on when you give a signal. Suggest they describe to themselves the action they are making as they create each sound. Where an instrument is blown, encourage the children to clean it with an antibacterial wipe before using it.

Recording
Give the children photocopiable page 33 which illustrates a range of instruments, some of which might not be within the child's classroom experience. Ask them to choose the correct names for the instruments and to describe the various actions they think are needed to create a sound with them. Some children can draw and describe other instruments on the reverse of the page.

Differentiation
Children:
■ explore sounds made by instruments; are aware of different ways of making sounds
■ explore sounds made by instruments, identifying different ways used to make sounds
■ show musical ability when exploring sounds made by instruments, identifying different ways used to make sounds.

Plenary
Point out that there are several different ways of making sounds when using instruments, which the children have tried out for themselves. Recall the different ways the children used the instruments to make sounds. Do they have a favourite method?

Display
Label and arrange the class instruments. Use labelled pictures of other instruments to add to these, particularly instruments from different countries or cultures.

SCIENCE

MUSIC

ACTIVITY 3

COMPARING WAYS OF MAKING SOUNDS

Enjoying sounds

© Ingram Publishing

Learning objective
To present and interpret results.

Resources
The collection of items from Activity 1 of this section and the instruments from Activity 2; pictures of other instruments or items that make sounds in different ways; a large sheet of paper; A4 paper; pens and pencils.

Preparation
Decide how the results of the exploration will be represented. The instruments and objects could be physically grouped and labelled in a prominent place for the children to discuss. Alternatively, or in addition, help the children make a large chart to represent the results.

This could be a simple column graph or a pictorial representation of the groups.

Provide a framework of boxes on an A4 sheet of paper for the children to make individual recordings. The boxes will represent the groups in which the children can draw and label the different ways of making sounds. Add questions or use sentence starters that will help the children to interpret the results of their exploration.

Activity
Recall the children's exploration of the simple musical instruments from the previous activity, pointing out that they have discovered there are different ways of making sounds using instruments. Remind them also of their investigations when making sounds with different objects (Activity 1). Suggest the children help you to group the instruments and objects according to the action that was required to make a sound.

Choose one instrument and ask the children how it can be used to make a sound. Then ask the children to identify any other instruments and items from the collection that are used in the same way. Put these together as a group. Repeat this for each method of creating sounds – blowing, tapping or banging, beating, shaking, plucking. Encourage the children to discuss their ideas, as the instruments and objects are grouped. Ask them to think of names for the groups. Is banging and tapping the same sort of action? Should things that require hitting, banging, tapping or beating belong to the same group? Are there any instruments that fit into more than one category? Where will they decide to put these instruments? If necessary, use pictures or photographs of instruments to add to the collection you have in the classroom.

When the sorting is complete, together look at the groups. What comments would the children like to make? Is there an obvious difference between the groups? Perhaps one group is much bigger than the others. Maybe the items in each group will need to be counted before any decisions can be made. If there is a group with very few examples, invite the children to think of an instrument or item that could be added. Ask them to decide upon the names for the groups so that you can make temporary labels.

Help the children to make statements about the groups. For example, *Most of the instruments and objects are banged or tapped to make a sound* or *We have only one example of something that you need to blow to make a sound*. Suggest the children help to make a chart to compare how many items there are in each group.

3

Enjoying sounds

Recording
Show the children the framework you have prepared on which they can present their results. Provide instructions for completing their recording. Point out that they are using the results of their exploration, which was to find out the different ways in which everyday objects and musical instruments can be used to make sounds.

Differentiation
Children:
■ group instruments and objects and with help compare how many there are in each group
■ group instruments and objects, making comparisons; help with presenting and interpreting the results
■ group instruments and objects, making comparisons, confidently helping to present and interpret the results.

Plenary
Point out the usefulness of sorting the items and how the groups and/or chart help to show what has been discovered. Ask the children to tell you what they have discovered from the results of their exploration.

Display
Leave the instruments and items in their groups. Print out labels with the names the children chose to describe each group.

SCIENCE   MUSIC

ACTIVITY 4

RECOGNISING SOUNDS

Learning objective
To identify sound sources and describe how sounds are being made.

Resources
Instruments and items from the previous activities in this section; a large piece of card that can be folded into three parts to make a simple screen to rest on a table top.

Preparation
Arrange a small screen on a table, behind which you can use instruments and other items to make sounds unseen by the children.

Activity
Tell the children that they are going to play a game. Behind a screen you have some instruments and items that you will use to make sounds. Tell the children you want them to try and identify what is making each sound. Go behind the screen and, for the first example, choose an easily identifiable sound source. Make a sound with it. Ask the class what they think could be making the sound. How did they recognise it so easily? Explain that they must have heard the sound before and remembered how it was made, perhaps during previous activities. Give individuals the opportunity to choose an item from behind the screen and to demonstrate the sound it makes for the class to identify.

Develop the game so that the children consider the differences in the sounds. Play one of the sounds again, perhaps louder or quieter, faster or slower. Could they identify what was making the sound when it was played differently? Were you able to trick them with any of the examples? Encourage the children to focus on the differences and to discuss them. Ask

some children to make sounds for the other children to identify and to describe how they are being played.

Extend the game by introducing an unfamiliar instrument or different item. Do the children find it as easy to identify this new sound source? Perhaps when playing another time, ask the children to provide a sound source of their own for others to identify.

Differentiation
Children:
■ take part in a game by identifying what is making a sound and describing how sounds are being played
■ take part in a game, identifying and discussing what is making a sound and describe how the sound is being played
■ showing musical aptitude when taking part in a game to identify sound sources and describe how the sound is being played.

Plenary
Talk about how enjoyable the game was and whether the children found it easy or hard to identify the instruments and items that were making the sounds. Suggest that as the children had heard these sounds during previous activities they had remembered them. Point out that it was more difficult to identify sounds they might not be familiar with because they had not heard them before.

Display
Make the game available for the children to play at specified times.

SCIENCE   MUSIC

ACTIVITY 5

SOUNDS AND NOISES

Learning objective
To recognise pleasant and unpleasant sounds.

Resources
A collection of sound sources, including instruments; pictures to represent pleasant and less pleasant sounds; board and writing materials; paper; pens, pencils and crayons.

Activity
Comment on the range of different sounds that the children have been making in the previous activities and sections. Make a quick list of words to describe sounds the children

have recognised. Include *loud, soft, high, low, all the time, now and again,* as well as *rattling, ringing, humming, tapping, scratching, crashing,* and so on. Suggest that some sounds have been enjoyable to listen to, but others not so pleasant. Ask individual children to tell you which particular sound they have liked. Suggest they select the instrument or item to demonstrate the sound they like. Encourage them to describe the sound and say why they like it. Do other people agree? Does everyone like this sound?

Move on to sounds that are perhaps not so enjoyable. What is it that makes them unpleasant? Ask some children to choose an instrument or item and demonstrate a sound they do not enjoy. Suggest that the children could call these sounds *noises*. Explain that noises and sounds are really the same, but we often use the words *noise* and *noises* when we do not like the sound and would perhaps like it to stop. Show pictures of sound sources and discuss whether the children think the sounds they make are pleasant sounds or unpleasant noises.

Ask the children to think of three sounds they like and three they do not like. Perhaps arrange the children into pairs so that they discuss each other's likes and dislikes relating to sounds. Suggest they think of sounds that they hear at home and in the streets on their way to and from school.

Bring the children together and ask for the sounds they like. Make a list of these on the board. Do the sounds have anything in common? Perhaps they are all quiet sounds. Are there any which are really loud? Are there both high and low sounds on the list? Are jingly sounds ones the children like to hear? Perhaps there is a whole range of sounds people like because everyone has a different favourite.

Focus on the sounds the children do not like. Make a second list on the board and discuss the examples. Perhaps very loud sounds are unpleasant. Some people might not like very high or very low sounds, or maybe scratching or rattling sounds. Loud sounds which go on for a long time are often unpleasant. Make some generalisations relating to the sounds the children do not like. Suggest there are sounds we like and dislike depending on how we feel. If we are trying to sleep or not feeling very well we might only like soft sounds.

Recording

Ask the children to describe the three sounds they like to hear and the three they dislike. They can illustrate the sound sources and use appropriate words to describe each sound. Perhaps provide the children with crayons and encourage them to use pattern and shape to represent contrasting sounds.

Differentiation

Children:
■ recognise that there are sounds they like and those they dislike; record using drawings and appropriate vocabulary
■ distinguish between pleasant and unpleasant sounds; describe using drawings and appropriate vocabulary
■ can describe why some sounds are pleasant and others unpleasant using drawings and relevant vocabulary.

Plenary

Remind the children that noises are sounds but often those we tend not to like. Point out that there are sounds we like and sounds we dislike, although these are not necessarily the same for everyone.

Display

Group the pictures, children's illustrations and other representations of sounds they like and dislike. Perhaps have an area to display *Sounds we like* and *Sounds we do not like*.

Using instruments to make sounds

What are these instruments called?
How do you play them?

_____

_____

_____

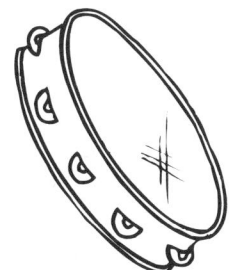

_____

_____

_____

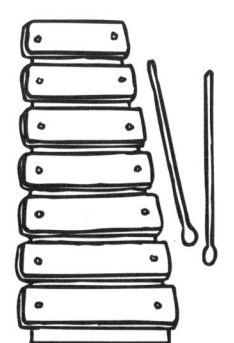

_____

_____

_____

_____

_____

_____

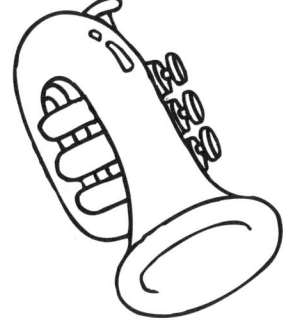

_____

_____

_____

_____

_____

_____

Instruments	Actions
trumpet tambourine drum guitar xylophone triangle	shake bang beat blow pluck tap scrape

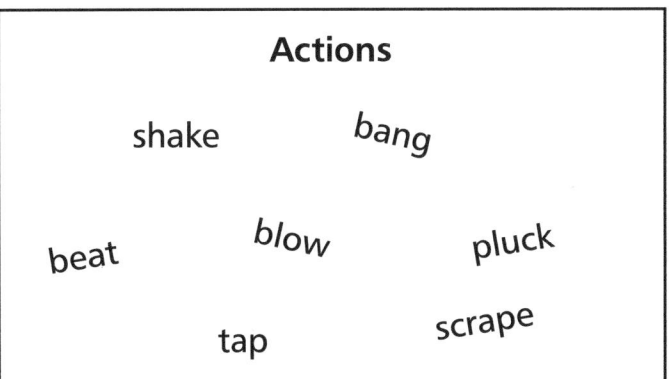

SCHOLASTIC

How we hear

SCIENCE

FOCUS
SCIENCE
■ understanding that the ears are sense organs
■ recognising that the ears provide information about the direction of sounds
■ raising questions relating to hearing
■ making and recording observations
■ using hearing for different purposes
■ understanding that sounds can be used as warnings, especially relating to road safety
■ understanding that sounds are louder closer to the source

ACTIVITY 1

SCIENCE

WE HEAR WITH OUR EARS

Learning objective
To know that we hear with our ears.

Resources
A simple, quiet sound source; photocopiable page 42; pens and pencils.

Preparation
The development of this activity will depend on the children's previous work relating to the senses.

© 2004/Photofusuion Picture Library/Alamy

Activity
Begin with a quick response session by asking the children to tell you what sounds they can hear as they sit in the classroom. These might include voices, chairs moving, a door opening, distant sounds from the corridor, the next room, the street outside, and so on. Ask the children how they are able to hear these sounds which tell them what is happening around them. They will talk about sounds reaching their ears. Ask if anyone can explain how they know that it is their ears that enable them to hear. Listen to the children's ideas and encourage discussion. The children will probably want to tell you that by covering their ears they do not hear so well.

Tell the children that you want to do a simple experiment to prove that it really is their ears that they are using to hear sounds. Demonstrate a sound, perhaps a tapping or the ringing of a bell. Ask the children to close or cover their eyes while you make the sound again. Was there any difference in the sound? Did they hear it as well? What does this tell us? Suggest that if we can hear sounds just as well with our eyes closed, we have shown that it is not our eyes which have anything to do with our sense of hearing.

Next, ask the children to cover their mouths while you make the sound. Continue the discussion. If appropriate, refer to the tongue as the sense organ enabling us to taste. Can the children explain why covering their mouths does not affect their hearing? Perhaps the children can tell you that next they should hold their noses, as they do when trying to avoid smelling something unpleasant, to see if this affects their ability to hear. Suggest this if they don't, and discuss again whether their sense of hearing is affected or not. (If the children want to include the sense of touch, ask for their suggestions as to how they might prove that their fingers are not involved in helping them to hear.)

Finally, ask the children to put their hands over their ears. Then ask them to describe what they heard. Did they hear the sound differently? Perhaps the sound was very faint. Perhaps they could not hear any sound at all. Help the children to understand the significance of this test, and how the outcome has proved to them that it is their ears and only their ears that enable them to hear.

Recording
Present the children with photocopiable page 42, which shows the outline of a child's head. Ask the children to make this head represent themselves by adding hair, colouring the eyes in, and so on. Around the picture they can illustrate and name the sources of the sounds they can hear as they sit in the classroom. Suggest they show how they hear the sounds by drawing lines from the sources to the correct part of their head.

Differentiation
Children:
■ know that they hear sounds through their ears, demonstrating this in a simple diagram
■ understand that they hear sounds through their ears, know how they can prove this and record it with illustrations and labels
■ understand the significance of showing in a scientific way that they hear sounds through their ears; record with illustrations and labels.

Plenary
Point out to the children that they have taken part in a scientific test which shows that it is through their ears that they hear sounds. They have answered the question *How do we know that we use our ears to hear?* Remind the children that animals other than humans have ears. Suggest the children collect pictures or make drawings of animals with unusual ears. They can perhaps do research at home and present their work as a booklet or poster.

Display
Collect pictures and photographs showing different animals' ears. Add any illustrations and notes produced by the children's research.

4

How we
hear

ACTIVITY 2

SCIENCE

LOCATING SOUNDS

Learning objective
To recognise that our ears provide information about the direction from which sounds come.

Resources
A simple sound source, such as a small bell; paper; pens, pencils and crayons.

Preparation
If the children are arranged into two groups they can observe the other group's responses. Make arrangements to take the children out of doors for part of the activity.

Activity
Position the children in two groups, perhaps sitting on the floor. Ask one group to face outwards and listen carefully. Tell them that someone, yourself or a child from the second group, will make a sound from somewhere within the classroom. Show them the sound source, which could be a small bell. Explain to the listening group that they must cover their eyes while the sound is made and then when they hear the sound, they should point to where they think it is coming from. Make sounds from different parts of the classroom for the children to locate. Repeat the activity with the second group, asking them to listen and locate the sounds.

  Ask the children what they have discovered from this simple experiment. They should be able to explain that their ears told them where the sound was coming from. They did not need to see the source of the sound because their ears provided the information.

  Take the children into the playground to identify where any sounds they can hear are coming from. First pick out a distinctive sound and ask the children where they think the sound source is located. Ask them to point to or face the direction of the sound source. Then identify other sounds and ask questions to stimulate discussion. For example, *I can hear a sound in this direction. What could be making the sound? What does this tell us? Do we think the sound source is close by or far away? What clues do we have?*

  Emphasise that although the children are not able to see some of the sound sources, they have found out what is happening nearby. It is their ears that have enabled them to hear and discover the information.

  Back in the classroom, make some statements relating to the discoveries made while listening. For example, *There is a lot of traffic in the town today. A tractor is working in a field behind the houses. There were birds in the branches of the big tree.*

Recording
Ask the children to make a drawing to show what they know is happening close by, even though they were not able to see. This can be a picture of the locality, illustrating and labelling the sound sources they heard. Some children might be able to indicate whether they think the sounds were quite close by or far away. The children can use speech bubbles to explain what they have heard.

Differentiation

Children:

■ are aware that their ears give them information about the direction from which sounds come, making simple drawings to show this

■ recognise that their ears provide information about the direction from which sounds come, with help recording what they have discovered using drawings and labels

■ recognise that their ears provide information about the direction from which sounds come, recording their ideas scientifically with drawings, notes and labels.

Plenary

Pick out a sound and ask the children where it is coming from. Then ask how they can tell. Point out that even with their eyes closed they can tell where sounds are coming from because it is their ears which provide them with clues. In this way they can discover things that are happening around them which they cannot see.

Display

If appropriate, reproduce and enlarge one of the children's drawings. Depict a small figure surrounded by buildings or trees listening to out-of-sight sound sources. A speech bubble can represent what the child can hear.

ACTIVITY 3

ASKING QUESTIONS

SCIENCE

Learning objective

To turn ideas about hearing into questions that can be tested.

Resources

Sets of earmuffs or ear protectors; a camera; framework for the children to record their results on; pens and pencils.

Preparation

Prepare a simple framework (see example, right) on which the children can record the progress of their test.

The children will write their names in the appropriate boxes as they take turns to complete each part of the test.

	Test 1	Test 2	Test 3
Organise the test	Ian	Holly	Glen
Make the sound	Glen	Ian	Holly
Listen	Holly	Glen	Ian

Activity

Remind the children of the previous activities in this section, where they showed that it is their ears that help them to hear sounds and their ears that also tell them where sounds are coming from. Point out that they have taken part in simple scientific tests to find answers to questions.

Encourage the children to suggest further questions about hearing which could be answered with a simple test. Guide the children towards thinking about why we have two ears. Do they think they can hear just as well with one? How could they find out if there is any difference using one ear or both ears? The children will perhaps suggest covering one

4

How we
hear

ear and listening to a sound, then maybe covering both ears and making comparisons. Guide the discussion so that the children devise a simple test to answer their question. Include the following points:

■ The test must be safe. Ears should be covered using earmuffs or ear protectors. Tell the children that it is dangerous to put objects in their ears (unless they are specially designed earplugs).

■ The test must be fair. The sound must be the same loudness each time and made in the same place.

Demonstrate the children's ideas for the test with the whole class. Let one child listen with earmuffs on one ear and then two, while another child makes a sound. Let the children decide what would be a suitable sound source to use.

Then arrange the children into groups of three and encourage them to set up their own test to discover any differences when one or both ears are covered. In turns, the children can organise the test, make the sound and experience the listening. Some children will need help to set up a simple test, others might want to make several trials and collect a range of evidence. If appropriate, provide the children with a simple framework on which they can record the procedure of the test. This will help them to follow a sequence and work in a scientific manner. They should write their names on their charts and add comments and observations. Take photographs of the children performing their tasks.

Discuss the children's observations as a class. How do the findings of the groups compare? Have they all come to the same conclusions? Where there are anomalies, try to find out what has happened. Are the findings what the children expected? Is anything surprising? At the end of the tests, can the children explain why they have two ears and answer the question?

Recording
Ask the children to draw labelled diagrams to show how they carried out their tests and what they have discovered. Some children can write the question they asked themselves and provide the answer. Photocopy the recording charts so that each child in the groups can have a copy.

Differentiation
Children:
■ with help suggest a question to test, make observations and present these as drawings
■ provide ideas for testing, take part in a test, making and recording observations
■ provide ideas for testing, organise and take part in a test, making and recording observations scientifically.

Plenary
Point out that the children have taken part in a scientific test. They have provided an idea for testing in the form of a question, organised the test safely and fairly, made observations and made a record of what they have discovered. Briefly talk about the results of the test.

Display
Use the recording charts, children's work and any photographs of the children performing their tasks to highlight the test.

ACTIVITY 4

SCIENCE

RECOGNISING HAZARDS

Learning objective
To understand that we use our sense of hearing for a range of purposes, including recognising hazards and risks.

Resources
A story, poem or anecdote which illustrates the importance of hearing in a dangerous situation; a video clip relating to road safety issues; a board and writing materials; paper; pencils and crayons.

Preparation
Refer to the school's Road Safety Policy, and make links and connections to fit in with any ongoing work on road safety.

Activity
Tell the children a short story or describe an incident where someone was alerted to a dangerous situation by a sound. This could be an explorer encountering a wild animal in a jungle, a person calling to warn someone of imminent danger or a pedestrian avoiding a noisy and dangerous vehicle. Talk about hazards that can be heard but not immediately observed and emphasise the importance of hearing in such situations. Discuss situations where people alert others of dangers by shouting a warning. Perhaps an old lady cannot see the hole in the road, a family does not realise their house is on fire, a fisherman is unaware that the river is so deep. Consider also how people can attract attention when they need help. For example, a climber trapped on a cliff edge, someone accidentally locked in a room, a person who is injured and cannot move. Point out that in these instances using our sense of sight is not enough and that our sense of hearing becomes very important.

Collect other examples. If the children are familiar with pantomimes, talk about the audience calling out *He's behind you!* at appropriate moments. Talk about the significance of a fire alarm and a whistle or bell at playtimes.

Show the children a video relating to road safety. Use this opportunity to emphasise the importance of listening, as well as looking, when out and about. On the board, list each of the noisy hazards shown in the video. Point out that pedestrians cannot look in all directions at once, but their ears are picking up sounds from all around. Although we cannot see around corners, our ears can alert us to the sounds of approaching danger.

Play the video again and look for dangerous situations, especially those where listening carefully is required. Discuss crossing roads and the need to listen, as well as to look about, for traffic. Point out that both our senses of seeing and hearing are important for keeping safe on the roads. Talk about the differences between town and country traffic, that it is just as important to listen in quiet places and busy spots. Emphasise that the children need to

4

How we
hear

be alert at all times where there is traffic, and that sometimes they might not be using their sense of hearing effectively. Talk about concentrating and making sure that sounds can reach their ears, especially when it is winter and their ears might be covered with clothing. Talk about how police cars, ambulances and fire engines alert people to their presence so that traffic can move out of the way to let them pass.

Briefly talk about how some animals use their sense of hearing to detect danger. Many animals can hear sounds much better than humans as they need to keep away from enemies. Even quiet sounds frighten birds, and dogs especially are always on the alert because of their exceptionally good sense of hearing. Suggest the children watch any pet dogs that are resting and notice how alert their ears are.

Recording

Use this opportunity to reinforce road safety messages. Ask the children to write out instructions for crossing roads, to include the importance of listening. Alternatively, the children can draw a scene illustrating several dangerous situations where people are warned of danger by sounds, such as people shouting or vehicles and machinery making loud noises.

Differentiation
Children:
■ are aware that their sense of hearing can alert them to danger, especially relating to road safety
■ are aware that their sense of hearing is used for a range of purposes, including warning of dangers, especially those relating to road safety
■ understand that their sense of hearing is used for different purposes and can provide a range of examples, including those relating to road safety.

Plenary

Talk about how valuable our sense of hearing is; that it enables us to enjoy many different sounds, to communicate with each other, as well as warning us of dangers we might not be able to see.

Display

Use road safety material to highlight the importance of using ears to detect danger.

ACTIVITY 5

LOUD SOUNDS

SCIENCE

Learning objectives
To know that some sounds can be heard from a long distance; that sounds seem louder the nearer you are to the source.

Resources
Ear protectors; board and writing materials; pictures, books or comics which show examples of loud noises and how these are represented in pictures; paper; pencils and crayons.

Activity
Talk about sounds that the children might have heard that they know are coming from a source far away. Make a list on the board. Include sirens from police vehicles and ambulances, thunder, people shouting, a burglar alarm or car alarm, the sound of an aeroplane overhead, a distant train, the hum of traffic on a motorway and perhaps the sound of the sea. Suggest

that the reason they can hear these far away sounds is that they must be very loud.

Ask the children if they have heard any of these sounds up close. Can they describe how the sounds are different when the source is near to their ears? Talk about the loudness of sounds. Is it pleasant to be close to any of these loud sounds? Do the children tend to cover their ears with their hands to lessen the noise? Which is the loudest sound they think they have heard? It might be a train passing at a level crossing, an alarm which goes off close by, noisy machinery at the roadworks, the engine of a plane or ferry. Make comparisons, such as the sound of an aeroplane's engine when it is far away, overhead and when nearby at the airport, the increasing noise of the siren when a police car approaches. Encourage the children to share their personal experiences. Perhaps they can describe what it is like walking by the sea when waves are crashing onto the rocks, when the radio or television is on very loud or when a baby is crying.

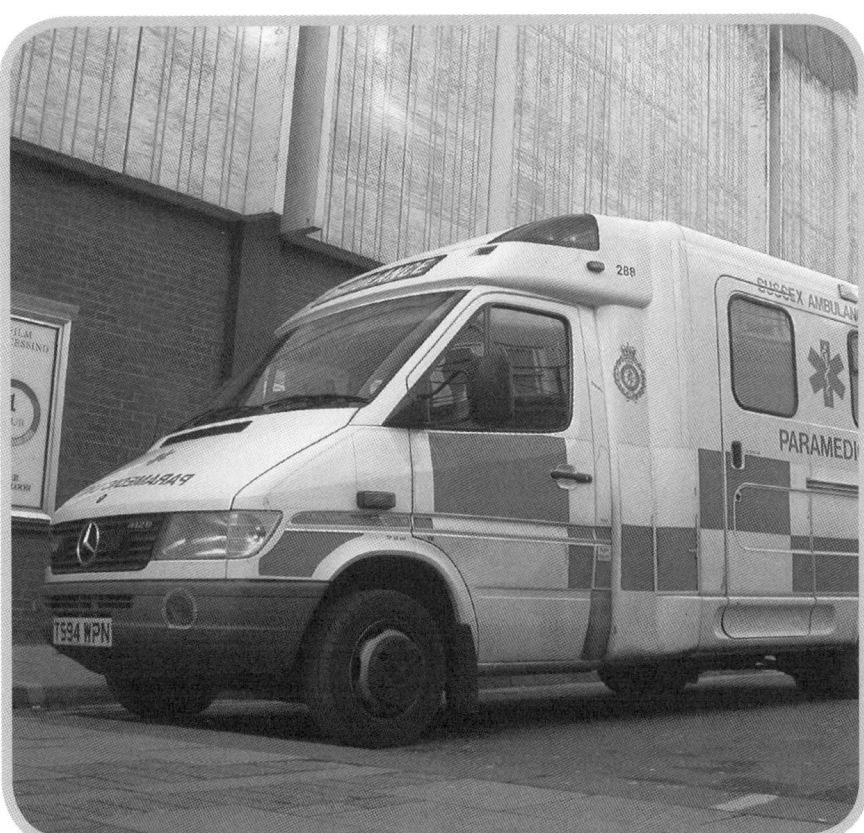

Tell the children that very loud sounds can harm their ears, especially if the sound continues for a long time. Talk about people who work in noisy situations. Ask the children if they know how these people protect their ears. Show them some ear protectors and find out if the children have seen people wearing them. Suggest road workers, tractor drivers and machinery operators. Advise the children against exposing their ears to continuous loud sounds.

Recording
Ask the children to illustrate some sources of loud sounds. They can include people working in noisy situations and wearing ear protectors. Discuss how they might show loud sounds in a drawing. Show them illustrations from stories, comics and so on, that do this.

Differentiation
Children:
■ recognise some sources of loud sounds; are aware that sounds are louder when close by and can be harmful
■ identify some very loud sounds, recognising that these are louder when close by and can be harmful
■ identify some very loud sounds, understanding that they are louder when close to the source and why they can be harmful.

Plenary
Ask the children how they feel about hearing loud sounds. Why is it unpleasant to be close to something that makes a loud noise? Make sure the children are aware that loud sounds can harm their ears.

Display
Create a 'loud' display. Use symbols and patterns, as used in comics and books, to indicate loud noises.

We hear with our ears

Investigation

FOCUS
SCIENCE
- carrying out an investigation
- comparing loud and faint sounds
- raising questions and making a plan
- making and recording measurements
- presenting results
- making comparisons and drawing conclusions

SCIENCE

Preparation for Section 5
Plan the procedure of the investigation. Decide upon a simple, workable test to demonstrate that a sound gets fainter as it travels away from its source. This could involve children listening to different sounds over one or more distances. A large area is needed in which to investigate. For example, an outdoor space or the school hall.

ACTIVITY 1

LOUD AND FAINT SOUNDS

SCIENCE

Learning objective
To know that sounds get fainter as they travel away from a source.

Resources
A small model of a fire engine or police car to represent a sound source and a figure to represent a person; a board and writing materials; paper; pens and pencils.

Preparation
Arrange to take the children out of doors for part of this activity.

Activity
Remind the children of your discussions relating to loud sounds and recall the dangers of being close to very loud sounds (Section 4, Activity 5).

Ask the children what happens to a sound when they hurry away from a very noisy engine or machine? Can the children explain that they would expect the sound to get fainter as they move further away? Ask the children to tell you what would happen to the sound if they were standing close to

5

Investigation

a fire engine with a loud siren as it rushed past them and away to a fire. They should expect the sound to become fainter until it disappears altogether. Point out that this time they are thinking about the sound source moving away. Ask some children to help you use the models of the fire engine or police car and the person to create these scenarios. Let other children provide explanations.

Explain that a faint sound is a soft or quiet sound that can seem to come from far away. Point out that it is easy for the children to hear your voice in the classroom. Point out, too, that they could hear your voice if they were standing next to you when outside, but what would happen if you moved away to the far end of the playground? Why would they be unable to hear your normal speaking voice? The children should be able to tell you that the sound of your voice would get fainter and fainter as you moved further and further away, until they could no longer hear your voice at all.

On the board write out and group vocabulary relating to loud sounds, including *loud, noisy, nearby, shouting, close to*, and soft sounds, such as *soft, quiet, faint, far away, in the distance*. Ask the children for their suggestions.

Take the children onto the playground or field to demonstrate faint and louder sounds. Choose an area where sounds can be demonstrated over a long distance. Divide the class into two groups. Ask group A to position themselves halfway along the distance and to chat to each other in normal voices when they get there. Can they be heard by group B who have remained behind? Ask the listeners to describe the sounds. Do they notice any difference from their own conversation? Then ask group A to move to the far end of the playground or field. Can group B still hear their voices? How would Group B describe any sounds they can hear? Help them to make comparisons. Reverse the groups and repeat the test. Point out that as the sound source moves away from the listeners it becomes fainter.

Perhaps try the test again, this time with the listeners moving away from the sound to show that as the source is left behind the sound gets fainter.

While out of doors ask the children to listen carefully for any faint sounds. Do they think the sounds are coming from far away? Can they identify the source of any faint sound they can hear? Is it easy or difficult to pick out faint sounds? Perhaps there are loud sounds close by that prevent them from hearing any faint far away sounds.

Recording
Ask the children to illustrate the listening test in which they took part. Suggest they draw a plan with simple figures to represent themselves in three different places when listening to sounds. At each point they can write their comments and observations of the sounds they heard. Encourage them to use relevant vocabulary for headings and labels, using the words on the board to help them if appropriate. Some children can write sentences to make comparisons.

Differentiation
Children:
■ are aware that far away sounds appear fainter than the same sounds made close by
■ know that sound becomes fainter the further the sound travels from its source, recording this with drawings and notes
■ know that sound becomes fainter the further the sound travels from its source, demonstrating their understanding with drawings and notes.

Plenary
Emphasise the discoveries the children have made by having taken part in a test. Point out that they have shown that sounds get fainter as they travel away from their source.

Display
Reserve a special area to display the investigation, possibly creating a bulletin board which links the stages of the investigation. Begin by arranging the vocabulary recorded in the activity, perhaps as a border.

PLANNING THE INVESTIGATION

SCIENCE

Learning objective
To encourage ideas and questions that could be investigated.

Resources
A prepared framework for planning the investigation; writing materials.

Preparation
On a large sheet of paper, prepare a framework of the plan for the investigation. This could be a set of questions to which the children will provide answers, such as:
■ What are we trying to show? What question are we trying to answer?
■ How will we carry out the test?
■ Where will the test take place?
■ What will we use to make the sound(s)?
■ How will we take measurements?
■ How will we make sure the test is fair?

Activity
Encourage the children to suggest ideas for an investigation to test how far away a sound can be heard. Recall the experiment from the previous activity, and suggest they should carry out a more scientific investigation that will provide results (evidence) which they will be able to show to others. Use questions and prompts to encourage the children to make suggestions and work out the details of the test. This should involve making a sound, making observations and measuring distances. Perhaps comparisons between two or more different sounds can be made.

Consider the following points with the children:
■ What are we trying to show? (That sounds get fainter as they travel away from a source.)
■ How will we carry out the test? (Work out the practical details of the test.)
■ What can be used to make the sound? (If a bell or drum is used, how can the children be sure they will always ring or bang it in the same way? Would a buzzer provide a more efficient way of providing sound for the investigation? If two sounds are to be used will they provide sufficient contrast? Perhaps a quiet sound and a louder sound should be tried?)
■ Where will each sound be made? (Perhaps at the far end of the field and halfway along.)
■ What will they use to measure distances? How will the measurements be recorded?
■ Will the person listening move away from the sound until it can no longer be heard, or will the sound be moved away from the listener?
■ What will they do to make sure that the test is carried out in a fair way so that the results will be of value?
■ What problems might be encountered? (Could there be other sounds getting in the way?)

Emphasise that it is very important to discuss the plan of the test in detail so that any problems can be sorted out and the test will go as smoothly as possible. Everyone who is taking part needs to know what is happening.

© Ingram Publishing

5

Investigation

Recording
With the children, write out a plan for the investigation by answering the previously prepared questions (see 'Preparation'). Ask the children to help you find the appropriate words while encouraging a scientific way of recording. Explain how useful it will be to have a written plan when the test is carried out.

Differentiation
Children:
■ with prompting, help to provide ideas for a class investigation
■ provide useful ideas and questions when planning a class investigation
■ begin to understand the importance of planning an investigation, provide ideas and questions relevant to a class investigation.

Plenary
Explain to the children that they are now ready to carry out the investigation they have planned. Because they have devised a plan, everyone will know what is happening and the procedure of the investigation is clear. Planning is an important part of any investigation.

Display
Display the plan so that the children can refer to it before and after the investigation.

ACTIVITY 3

CARRYING OUT THE INVESTIGATION

SCIENCE

Learning objective
To measure distances and record measurements as part of a class investigation.

Resources
Measuring/metre sticks; clipboards; a prepared recording framework for each group; pencils; a camera; video camera.

Preparation
Plan the procedure for the test. If two sounds are being tested, divide the class into four groups. Two groups can be involved with each sound. For each sound, one group can make the sound and do the measuring and recording, while the other group acts as the listeners. This should be repeated so that everyone has an opportunity to listen. Invite the waiting groups to observe and check that the procedures are followed correctly and scientifically.
   Prepare a framework on which the children can record the measurements.

Activity
Show the children the plan for the investigation from the previous activity and ask them to tell each other what they are going to find out by carrying out this investigation. Praise sensible behaviour, with all children focusing on the test. Point out that they are carrying out a test as part of a scientific investigation. Referring to the plan, inform them of the procedure for the test. Show the children the recording framework and explain how the results will be collected and recorded. Insist on accuracy and try to promote a scientific method of working. Explain that the results are important as they show what has been discovered.
   At appropriate points during the activity, ask the children to tell each other how they can show the test is being carried out in a fair way. Are they controlling the sound source in the same way for each part of the test? Are the children always facing the same way when

listening? Make sure the children check that each group uses the same units and takes measurements in the same way.

If appropriate, take photographs of the children carrying out the test. If a video camera is used the children can perhaps provide a commentary in their own words.

Recording

The children can write an account of the test, stating what they were trying to find out, what part they played in the test, how the results were collected and how they made sure the test was fair.

Differentiation

Children:
■ take part in a class investigation, helping with measuring and recording
■ take part in a class investigation that involves measuring and collecting results; work in a scientific manner
■ take part in a class investigation, understanding the importance of collecting and recording evidence and working in a scientific manner.

Plenary

Comment on the success of the test and the manner in which the children worked. Explain that working sensibly, collecting measurements and recording accurately, contribute to a scientific way of working.

Display

Include the children's work and add any photographs with labels.

ACTIVITY 4

PRESENTING THE RESULTS OF THE INVESTIGATION

SCIENCE

Learning objectives

To communicate what has been discovered using a chart to present results; to make comparisons between results and draw conclusions.

Resources

Materials to create a graph, such as a large sheet of squared paper for a block graph; crayons and pencils; computer software; the results from the test.

Preparation

Decide on the best method of presenting the results. If the measurements are appropriate a block graph can be used. Alternatively, the children could draw arrows against a simple scale to represent the distances. Arrange to make a class graph during the activity. If appropriate the children can also make individual graphs, perhaps using computer software.

Activity

Explain that the measurements the children collected during the tests in the previous activity are important. They are evidence to show what has been discovered. Talk about how these measurements could be presented to show others what has been found out. Point out that the information must be clear and easy to understand.

Depending on the children's experience they might suggest making a chart or a graph to present the results. Refer to any presentation methods they have used in the past. Encourage

Investigation

the children to discuss the form a graph would take. Ask questions, such as *How will you show the distances? Where will you represent the sound sources?* It might be helpful if they imagined the graph to be a plan of the area where the tests took place. With the children's help, draw the graph on the large sheet of paper. Add necessary information to the graph, such as a heading and labels.

Emphasise that the finished graph shows the results of the test and is an important part of the whole investigation. It shows others what has been discovered and can be used to complete the investigation, as it will help the children to make comparisons and conclusions.

Ask questions that will require the children to use the information contained in the graph and help them to compare the results. Link the listeners' experience of hearing the sound with the distance travelled. Ask, *Did each sound travel the same distance? Which sound travelled the furthest? Which travelled the shortest distance? How does the graph tell us this information? What were you using as an indicator to tell you how far the sound travelled? Are the results what you expected? Did the louder sound travel the furthest? How can we tell this from the graph?*

Encourage the children to talk about the information the graph provides. Explain that they need to show what they have found out by doing the investigation – that they have answered the question they suggested at the beginning of the investigation. Decide on a concluding statement, which might be *Sounds get fainter as they travel further away from where they were made.*

Recording

Some children might like to produce their own graph of the results. Alternatively, provide a copy of a computer generated-graph for the children to include with the rest of their recording relating to the investigation. Help the children to write a concluding statement to show what has been discovered as a result of the investigation.

Differentiation

Children:
■ help to create a graph and provide a concluding statement to complete the investigation
■ know that a graph is useful when presenting results, provide ideas for creating a graph, making comparisons and a concluding statement to complete the investigation
■ begin to understand the significance of a graph when presenting results, provide relevant ideas for creating a graph, making comparisons and a concluding statement when completing an investigation.

Plenary

Talk to the children about the stages of the investigation. Remind them that they began by asking questions, which is what all scientists do. Then a plan was needed before the test was carried out so that preparations could be made and everyone would know what they would be doing. The measurements were recorded accurately so that the results could be presented as evidence or proof for others to see. They did this by transferring their results to a graph. From the results the children were able to make comparisons and conclusions. Now they can provide the evidence to show the answer to their question.

Display

Add the measurements and graph(s) to the investigation display. Print out the concluding statement. Include a summary piece to explain to visitors what the children set out to do and how they achieved it.

Expressive sounds

MUSIC

FOCUS
MUSIC
- ■ identifying musical instruments
- ■ using musical instruments
- ■ playing instruments in different ways
- ■ creating sound effects
- ■ following visual instructions
- ■ practising control of instruments
- ■ selecting sounds to reflect moods

MUSIC

ACTIVITY 1

EXPLORING SOUNDS

Learning objective
To identify instruments and explore how they can be used.

Resources
A selection of simple instruments; card for instructions; a board or large sheet of paper and writing materials; A4 paper; coloured pens.

Preparation
Reserve some instruments for the second part of the activity. Write simple instructions on cards for making different sounds, such as *play quietly, make a rattling sound, make a loud sound, play fast but quietly, make a scraping sound*. Make a vocabulary list on the board or large sheet of paper to include *loud, quiet, tapping, banging, scraping, scratching, rattling, tinkling, jingling*.

Activity
Arrange the children so that they are sitting comfortably in a circle. Show them an instrument, such as a small drum, and ask them to tell you what it is called. Then ask them to describe the sounds they think it can make. After the discussion, ask an individual to make a sound using the instrument. Encourage the children to describe this sound – perhaps it is a loud bang, perhaps a quiet tapping. Hand the instrument to the next child and ask them to make a different sound. How can the children describe this new sound? Pass the instrument around the circle until the children have made and described as many different sounds as possible.

Produce a different instrument and follow the same procedure. From their experience of the first instrument the children should be able to make interesting sounds and provide descriptions with confidence. Talk about *banging, tapping, scratching, scraping, rattling*, and so on, referring the children to the vocabulary list. Encourage the children to include loud and quiet, fast and slow, continuous and intermittent sounds. Add any of the children's interesting descriptions to the vocabulary list. Continue so that each child has the opportunity to make several sounds and help with the descriptions.

Next, tell the children that you have some instructions for them to follow. Show them the

6

Expressive
sounds

cards and some different instruments. Ask a child to take a card, read the instructions and then choose an instrument with which to make the sound described. Can the rest of the class tell you what the instrument and the instruction was? Encourage the children to reply in sentences. For example, *Kay chose a triangle to make a tinkling sound.* If possible give each child the opportunity to interpret an instruction and make a sound.

Recording
Ask the children to draw and label three instruments and then describe the sounds they can make with these instruments. Refer them to the vocabulary list.

Differentiation
Children:
■ identify different instruments and explore and describe the sounds they make
■ identify different instruments and explore their sounds, providing appropriate descriptions and recording with drawings and words
■ identify different instruments and explore the sounds they make, offering accurate and imaginative descriptions, recording with drawings and words.

Plenary
Comment on the variety of sounds the children produced when using the instruments. Did they think there were so many ways of making sounds, or so many ways of using different instruments?

Display
Arrange the instruments and provide name labels for the children to match to them.

ACTIVITY 2

COMPARING SOUNDS

MUSIC

Learning objective
To play instruments in different ways and compare sounds.

Resources
A selection of instruments; cards with sound sources on; paper; pens, pencils and crayons; an electronic keyboard.

Preparation
Make cards with illustrations or words to show or describe some everyday sound sources, such as a doorbell, a bicycle bell, wind chimes, a car, a tree blowing in the wind, a waterfall, water splashing, and so on.

Activity
Arrange the children so that they are sitting in a circle around a collection of instruments and the cards. Ask a child to choose one of the cards randomly. When they have looked at their card, ask them to select an instrument from the collection which they think can make a similar sound to the sound source illustrated on the card. Encourage the child to explain their choice before demonstrating the sound. Is the sound what they expected? Do all the children think the sound is an accurate representation of that on the card? Has anyone a suggestion for a different instrument? If so, let them try making the sound and then ask the rest of the class to compare the two sounds.

Give other individuals the opportunity to make sounds in the same way.
Next, ask a child to choose a card without showing the rest of the class and then to select an instrument to make the mystery sound. Can the other children suggest what the sound represents? Is it easy or difficult to guess? The class may need help. Encourage further clues to enhance the image, perhaps the sound of wind to accompany the chimes or a car horn to add to the sound of the car engine. Maybe ask a second child to try making a sound with a different instrument.

Some children may be able to think of their own sounds to demonstrate for the class to guess. Let them try these, using the instruments and asking the class to guess what sound they are representing.

As the children's skills at this game develop, insist on more detailed interpretations and descriptions. For example *A person walking slowly, Water dripping quickly.*

Let the children work in pairs. Provide each pair with a single instrument and ask them to practise a sound they think is particularly good. Bring the class together and let each pair demonstrate their special sounds to the rest of the class. If appropriate, finish by using an electronic keyboard to demonstrate different ways of making sounds.

Recording
Ask the children to choose three different instruments to draw and to write the sounds they can imitate with them.

Differentiation
Children:
■ play instruments in different ways to make specific sounds
■ play instruments in different ways recalling past experiences to make specific sounds, providing descriptions and making comparisons
■ show imagination and flair when playing instruments and recalling past experience to make specific sounds; provide descriptions and comparisons of and between sounds.

Plenary
Comment on the range of sounds the children have made. Suggest the children continue to explore instruments, probably at home, to discover new ways of making sounds.

Display
Add name cards to any new instruments used in this activity.

ACTIVITY 3

FOLLOWING INSTRUCTIONS

MUSIC

Learning objective
To respond to visual signals when making music.

Preparation
Select songs for the children to sing and decide on hand signals to use to convey different instructions.

Activity
Choose a short, familiar song that the children like to sing. Explain that you are going to 'conduct' the singing using your hands, so as well as listening to the singing the children must watch carefully for any signals.

Encourage the children to sing the whole song as a trial run while you beat time in a simple manner. Next, show the children what the stop signal will be and how you will change from keeping the beat to this new signal. Ask the children to sing the song again but to be prepared to stop when they see the signal. Use the signal several times to test the children's reactions. At this stage of the activity, use a verbal instruction each time you want the children to begin singing again.

Next, develop a hand signal to tell the children when they should begin singing. Encourage the children to sing the whole song again without any verbal instruction, but watching carefully for visual signals. Start the singing and include several stops and starts during the song using hand signals only. You could perhaps tell the children you want to test their reactions by asking them to sing a different song while you conduct them.

Ask the children what other signals they might find useful when singing. They might suggest knowing when to sing loudly and when to sing quietly. Discuss possible signals and try out some of the ideas. For instance, if you have decided on signals to show loud and quiet, let the children sing again and add these signals to those that they already know.

Ask the children why they think visual signs are more useful than spoken signals when they are singing. Talk about not interrupting or spoiling the song, and point out that the children cannot tell each other what to do if they are using their voices for singing. Explain how a conductor can take charge of a song using visual signs.

Use all the learned signals in a favourite song. Perhaps individuals would enjoy conducting the singing using the visual signals. Provide opportunities for short bursts of singing at other times so that the children can continue to enjoy practising their reactions to the newly learned signals.

Differentiation
Children:
■ begin to respond to visual instructions when singing
■ respond to visual instructions when singing, becoming aware of the importance of these
■ respond accurately to a range of visual instructions when singing, understanding the value of these signals.

Plenary
Point out the importance of a conductor when a group of people is singing or playing instruments. Ask the children to remind each other what kind of things the singers or instrumentalists need to know. For example, when to begin, or when to sing or play loudly or softly. Ask the children to find out examples of when visual signals are useful in everyday life. People might use signals when they are far away or if it is too noisy to hear voices.

ACTIVITY 4

CONTROLLING INSTRUMENTS

MUSIC

Learning objective
To practise the control of instruments.

Resources
A selection of musical instruments.

Preparation
Select instruments so that the game starts with those that are easier to control and progresses to more difficult examples.

Activity

Arrange the children in a circle and tell them that you have a musical game for them to play. Explain that this game requires a very special skill. Produce an instrument, such as a small drum, and tell the children they will all have a turn with the drum. Tell them, however, that when they are in possession of the drum they must hold it carefully so that it does not make a sound. Then they must pass it on to the next person without it making a sound. Start the game with the drum going around the circle. Can the drum get back to the starting position without any sound being made? Encourage the children to act sensibly. Be sympathetic towards anyone who might drop the instrument or make an accidental sound. Praise the children for their efforts as appropriate. Ask if anyone has any tips for others as to the best way(s) of controlling this instrument.

Move on to other instruments, progressing to those that are more difficult to control, for instance a tambourine or a set of bells. Do the children think they are good at controlling instruments? Which instrument was the hardest to control? Was there one that most people had a problem with? Which was the easiest?

Extend the game by asking individuals to perform small tasks while holding an instrument. For example, *Stand up and sit down again*, *Walk across the room*, *Turn around twice*. Explain that they must not make a sound with the instrument.

Ask the children why they think it is important to be able to control an instrument and prevent it from making a sound. Talk about performing and playing only at the appropriate time when in an orchestra or group of musicians. Explain that remaining silent when in charge of an instrument is just as important as making the sounds.

Choose an instrument and demonstrate a simple sound pattern, perhaps two taps and a shake using a tambourine. Ask each of the children to follow this pattern, passing the instrument around the circle and trying not to make any other sounds in between. Repeat with a different instrument.

Differentiation

Children:
■ make an effort to control musical instruments
■ handle and play instruments with control
■ handle and play instruments confidently, understanding the importance of control.

Plenary

Point out how difficult it can be to control musical instruments and prevent them from making sounds at the wrong moments. Praise the children's efforts and ask them to remind each other why they need to practise their control of instruments.

Display

Arrange pictures of conductors with choirs and orchestras.

ACTIVITY 5

SINGING AND PLAYING

MUSIC

Learning objective

To make and select sounds to reflect the mood of a song using simple musical instruments.

Resources

A selection of musical instruments, including a keyboard if available; a recording sheet.

6

Expressive sounds

6

Expressive sounds

Preparation

Select familiar songs that the children can accompany by playing instruments. Choose contrasting songs where the same instrument can be used in different ways, or perhaps a song with several verses, each of which can be interpreted differently. For the children's recording, prepare a sheet with a centre box and a border of about 3cm.

Activity

Ask the children to sing a particular song that they already know. Use the visual symbols that the children understand from Activity 3 of this section to conduct them. For instance, instruct them to begin, and beat time while they sing. Then suggest the song could be made more interesting with a musical accompaniment. What sort of sounds do the children think would enhance the mood of the song? Lead them towards thinking about what type of sound the song might need. Is it a loud or a quiet song, a jolly or a sad song? Perhaps the song is about an animal that makes a distinctive noise, or perhaps the song includes the weather and the sounds of rain or thunder can be added.

When the children have decided what sort of sounds they want to add as an accompaniment, talk about how the sounds could be made and which instruments could be used. Encourage the children to draw on their previous experience and recall the sounds they have discovered during their explorations of instruments. Then ask some children to use instruments and make sounds that might be suitable for the mood of the song. Help the children to decide on which sound they think is the best. After a short practice, ask the children to perform the song again, this time with the musical accompaniment.

Repeat this process with a contrasting song. What are the children's suggestions for a sound accompaniment now? Can the same instrument be played in another way? Does a different instrument provide a sound to match the mood of the song? Provide the opportunity for half the class to perform while the rest listen and enjoy the music.

If appropriate, use a keyboard to demonstrate the different moods that can be created within a song or by contrasting the moods of the two songs.

Pairs or small groups of children could choose their own song and practise an accompaniment for the rest of the class to enjoy.

Comment on the importance of finding the most suitable sound to create the mood of the song. Point out how the same instrument can be used in quite different ways, perhaps to create gentle sounds as well as busy sounds. Talk about instruments that are best for creating quiet, sleepy moods, or those which usually make us feel busy and energetic.

Recording

Show the children the recording sheet. Ask them to illustrate a scene from one of the songs in the box in the centre. Tell them they should use the border to draw the instrument they used to accompany the song and write the sound it made. Suggest they could make an interesting repeating pattern for the border by alternating the instrument with its label.

Differentiation

Children:
■ enjoy helping to make and select sounds to create the mood of a song
■ enjoy recalling suitable sounds to play when creating the mood of a song
■ enjoy recalling and selecting suitable sounds to play thoughtfully and accurately, when creating the mood of a song.

Plenary

Point out that as well as using their voices to create the mood of a song, perhaps by singing slowly, quietly, loudly, and so on, instruments can be used in many ways to make a song more interesting. Anyone who is listening can enjoy the song and understand a little more about the message of the song.

A sound story

FOCUS

MUSIC
■ selecting sounds and sound sources to use to enhance a story or poem
■ using sounds expressively

SCIENCE
■ recalling past experience of using sounds

SCIENCE

MUSIC

ACTIVITY 1

LISTENING TO A STORY OR POEM

Learning objective
To listen to a story or poem and identify sound sources.

Resources
A story or poem with suitable sound sources in it; a board and writing materials; paper; pens, pencils and coloured pens.

Preparation
Choose a story or poem with a variety of sound sources in it. This could be a favourite story or poem which the children have heard many times, or a story or poem new to most children.

Activity
Read the chosen story or poem to the children. Ask them to tell you what they like about this particular story. If it is one they have heard before can they try to explain why they like to hear it over again? How do they feel while they are listening to the story? What sort of mood does it create? Do the children think it is a quiet, gentle story? Perhaps it is a noisy, busy story in which lots of things happen.

Ask the children to tell you any sounds they can remember from the story. What are the sources of the sounds? Maybe there are people or animals creating the sounds. Discuss whether the people or animals are using their voices or moving around noisily. Perhaps it is the weather or machinery that contributes to the sounds of the story.

Tell the children that you will read the story again, and ask them to put up their hands when they hear a sound described, or can tell something is making a noise. Make a list of the sounds on the board as the children identify them. Add ticks to show where a

7

A sound story

sound has been repeated. Depending on the content of the story it might be useful to group the sounds according to their source. For example, people sounds, animal sounds, vehicle sounds, weather sounds.

Remind the children of their search for different sounds made by voices (Section 2, Activity 2). Are there any words in the story that sound like noises when they are spoken? (Onomatopoeic words.) Has the writer cleverly used words that match the sounds they are describing? For instance, *crash*, *thump* or *screech*.

Ask the children to pick out some of their favourite parts of the story where sounds are effectively used. This could be where a character shrieks or an image of water is created.

Recording
The children can write their own account of the sounds they recognise in the story. Alternatively, provide them with sentence starters, such as *I think the loudest/softest sounds are… , Most of the sounds are made by… , The sounds I like best are…*
The children could illustrate incidents within the story where a sound is featured.

Differentiation
Children:
■ recognise and describe some sounds in a story or poem
■ recognise sounds in a story or poem, describing the overall sound picture created
■ recognise sounds in a story or poem, identifying words used to create sounds and describing the overall picture.

Plenary
Suggest that one reason why the story or poem is a favourite of the children's is because of the sounds they imagine within the content. Remind them of some special moments in the story or poem.

Display
Print out the poem or a part of the story and highlight the sounds featured. Perhaps Add any illustrations the children have made.

SCIENCE   MUSIC

ACTIVITY 2

DISCUSSING SOUNDS

Learning objective
To use knowledge and experience to identify sounds that can be made to enhance a story or poem.

Resources
The story or poem and list of sounds from the previous activity; a large sheet of paper; coloured pens; a display of instruments and items used for making sounds throughout this topic.

Activity
Suggest that the story or poem which the children enjoyed in the previous activity could be even more exciting and enjoyable if some extra sounds were added. Refer to the list of sounds in the story that the children helped to compile, and highlight any sounds they think they could make. Explain that these sounds would be made as the story is read and would make the story even more interesting.

Remind the children of their explorations with their voices, everyday items and musical instruments in previous activities. Show the children the musical instruments and items they have used in previous activities if appropriate.

Focus on five or six sounds that the children have suggested making from their list of sounds, and write down the children's ideas for making each one. For example, *Glen remembers that he made a really good sound like thunder using the big drum* or *Alice can show us how to make a sound like raindrops using her fingers on the table.* Try to distinguish between sounds that match a specific sound in the story and those in the background that help to create a mood. Build up a list and develop the discussion so that individuals feel they can make suggestions while accepting that their ideas might be changed. Point out that everyone is making a contribution and that the final decisions will be a mixture of all the ideas. Try to ensure some sounds are represented by instruments, some by classroom items and others by using voices in different ways. Include any handmade instruments if possible.

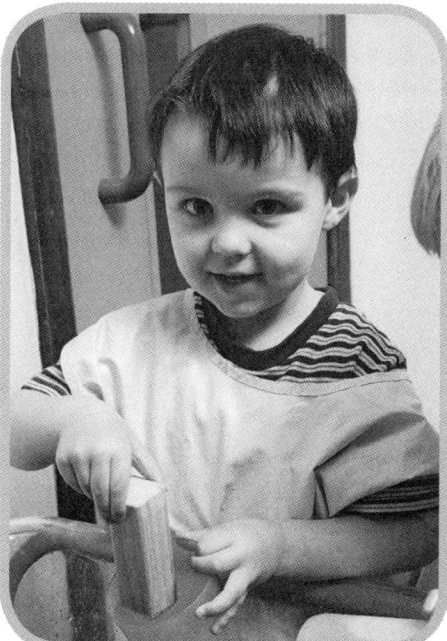

Summarise the list with the children. For example, *We have lots of ideas for... and just two suggestions to show...* Check where instruments might be used and underline or circle these with a coloured pen. Using different colours, do the same for everyday items and for voices.

Perhaps finish by reading the story again and ask the children to consider where the sounds should be made and imagine how they might fit in. Suggest that some sounds can be made where there is a pause, while others can be made alongside the reading.

Differentiation
Children:
■ with help provide suggestions for making sounds to enhance a story or poem
■ can draw on previous experience to provide ideas for making sounds to enhance a story or poem
■ can draw on previous experience to provide accurate and imaginative sounds to enhance a story or poem.

Plenary
Tell the children that you think their ideas will make the story more exciting and interesting to listen to. Point out that they have used good ideas, which they remembered from their explorations of different types of sounds.

Display
Display the list of ideas for the children to refer to and discuss.

SCIENCE   MUSIC

ACTIVITY 3

EXPERIMENTING WITH SOUNDS

Learning objective
To make sounds to match images and events in a story or poem.

Resources
Everyday items and musical instruments which the children have explored in previous activities; the list of sounds from the previous activities in this section.

7

A sound
story

Preparation
Arrange the items and instruments so that they can be selected when needed by the children.

Activity
Tell the children that now they should try out the sounds they have suggested to accompany the story and see which ones sound best. First, choose one of the sound suggestions that all the children can make with their voices or parts of their bodies. This might be a whistling sound or an animal sound made with their mouths, or a tapping or clicking sound made with their fingers. Try to find a sound that can be repeated several times during the story or which can be used as a background effect. Spend a short time letting the children practise this sound. Ask groups of children to make the sound while others listen to assess its effectiveness. Read a section of the story where the sound will be introduced and ask the children to try out the sound. Do they think it is appropriate? How could it be improved? Is it too loud, too soft, too fast, too slow?

Next, choose a sound for which there are several ideas. Suggest individuals demonstrate each idea while others decide which is suitable, or whether any changes need to be made to make one sound better than the others.

In pairs, the children can try out some of the other ideas for themselves. Explain a procedure for the practice sessions, suggesting ways of limiting the sound so that other pairs are not disturbed, and encouraging amicable discussion. Ask the pairs to find out if the sound they are trying out is effective and if they can suggest any improvements to it.

After a suitable practice time, bring the children together and ask the pairs to demonstrate their ideas. Find out who has a good idea for a particular sound. Does the rest of the class agree that this sound is exactly right? Has anyone a suggestion for an improvement? Is the sound loud or soft enough? Should it be slower or faster, higher or lower? Read the relevant part of the story each time a new sound is presented so that the children can assess its suitability.

Decide on all the sounds required to enhance the story. Read out the appropriate part of the story for each sound and ask the children to decide how long each sound needs to be made or played for. Discuss when sounds will be made by the whole class, by groups of children or by specific individuals.

Differentiation
Children:
■ explore sounds, trying out suggestions to enhance a story or poem
■ explore sounds and provide and try out useful suggestions to enhance a story or poem
■ explore sounds and make imaginative suggestions for enhancing a story or poem.

Plenary
Point out how useful it was to have explored so many different ways of making sounds. As a result the children have been able to provide ideas and suggestions for sounds to accompany the story.

SCIENCE MUSIC

ACTIVITY 4

REHEARSAL

Learning objective
To understand that practice is important when putting ideas together for a performance.

Resources
Items and instruments to make sounds, as used in the previous activity; a camera.

Preparation

If necessary, devise instructions for the children to follow which will tell them when to make their sounds. Alternatively, a short poem could be printed out with colour-coded information added to tell the children when they should make their sounds. For a longer poem or story use visual signals to help the children.

Activity

Explain that you are going to record the story with added sounds as the children perform it, so that they can listen to it and enjoy it again. First, however, the children will need to rehearse so that everyone will know what they are doing.

Go through the story and discuss where the sounds will be played. Make sure the children know when they are playing as a class, in a group or as an individual. Explain any help that the children will have to inform them when to play. Perhaps the children can use words from the story as cues and prompts. If there are colour-coded instructions point these out. Explain any visual signals you will give the children. Remind them of the importance of controlling all the sound sources so that sounds are only made when required. Can they remember the signal that means silence?

Ask the children to select their instruments and other sound sources. Perhaps some individual practice will be needed first. Invite the children to try out their sounds in turn. Practise the sounds that everyone will make together. If appropriate, take photographs of the children practising their sounds.

Arrange the children in a group so that they are comfortable, can make their sounds successfully and are able to see any visual signals you will be giving them. When you think they are ready for the rehearsal have all the children sitting quietly, anticipating their performance.

Read the story or poem with the children making their sounds. After the first run through, briefly discuss any problems. Perhaps someone played too loudly. Perhaps someone was unsure when to play. If necessary, repeat some or all of the practice. Suggest the children practise their particular sounds at other appropriate times, too, possibly when at home. Talk about the importance of practice before a big event.

Differentiation

Children:
■ with help take part in a rehearsal making group and individual sounds
■ take part in a rehearsal, making group and individual sounds when appropriate
■ take part in a rehearsal, competently making group and individual sounds as appropriate.

Plenary

Tell the children how well their rehearsal went. If necessary suggest any improvements in behaviour and technique. Emphasise the importance of being able to work together as a group to produce this kind of performance. Point out that musicians spend many hours rehearsing to make sure they will play their best for their audience.

Display

Display any photographs you took during the activity with captions. This will add importance to the children's efforts.

7

A sound story

SCIENCE

MUSIC

ACTIVITY 5

PERFORMANCE

Learning objectives
To take part in a performance, using sounds expressively to illustrate a story or poem; evaluating a performance.

Resources
Items and instruments from the two previous activities; recording equipment; paper; pens, pencils and crayons.

Preparation
Let the children know when the final performance and recording will take place so that they can look forward to the event.

Activity
Ask the children to collect the items and instruments they need for making their sounds. Arrange the children in a suitable group for the performance. Remind them that every sound they make will be picked up by the recording equipment, so they must remember to control their voices and their instruments once the performance has started. Suggest that everyone has a cough to clear their throats beforehand.

Record the performance as you read the story, with the children contributing as practised.

Congratulate the children on their efforts. Play the recording while they listen and enjoy their performance.

Discuss the performance. Which part did they enjoy the most? Which sounds do they think are particularly good? Could they hear all the instruments or items? Would they make any changes? Are they pleased with their efforts? Did they enjoy working towards the final performance? Does the performance seem different now they are listening to it rather than performing? Ask the children who else they think might like to hear their performance. Perhaps make arrangements to play it to another class.

Recording
Ask the children to write an account of their experience of taking part in the performance and include an illustration. Perhaps provide sentence starters, such as *The part I liked best...* or *When I listened to the recording...*

Differentiation
Children:
■ take part in a class performance and enjoy making sounds expressively
■ take part in a class performance, enjoying making sounds expressively and offering suggestions for improvements
■ show musical ability when taking part in a class performance, enjoying making sounds expressively and offering relevant suggestions for improvements.

Plenary
Praise the children for their efforts in choosing the sounds, in practising and rehearsing them. Point out they used different sounds to enhance a story which others will be able to enjoy.

Display
Arrange the children's writing and illustrations. Include information for visitors, explaining the children's part in the preparation and performance.

Display

It is important to display as much as possible of the children's work relating to sound. Making their efforts available for others to see and enjoy not only adds value and prestige to what the children have achieved, but also emphasises the importance of the information they have discovered, ideas they have had and explorations they have made.

If the children know their work is on show for others to examine and discuss, they will take care over presentation. At this stage the habit of always producing their best efforts should be encouraged.

Individual presentation

Collect each child's individual paper work – drawings, completed worksheets, written accounts and reports – and present them in a folder or as a booklet which they can personalise and show to each other and their parents and carers. If necessary, any pieces of work needed for classroom display can be photocopied.

Classroom display

It can be useful to display each child's work relating to any activity in a temporary way, perhaps with Blu-Tack, so that discussion and evaluation can take place and the work can be admired. At this stage, make sure the work is displayed on a level where it can easily be seen by all the children. If space is limited in the classroom, extend the displays into corridors or the school hall.

Wherever possible, encourage the children to help with the design and arrangement of the displays. Ask them to suggest suitable headings, help to choose colours and backgrounds and think of ideas for arranging their work.

Frequently refer to the displays as the topic develops, reminding the children of the different elements of their work and the progress they are making. Make links and connections across different areas of work.

It is important to emphasise the value of scientific observations and investigation by providing an area to display these aspects of the topic. This helps children to become aware of the importance of communicating results and conclusions to others.

As vocabulary is an important element of many of the activities, it is useful to collect groups of words and make these easily available for the children's use. Words can be displayed as lists, in a class book, photocopied and made into individual booklets and printed on cards that can be used for matching and attaching to displays in a temporary way.

Prepare an area where the musical instruments and other items used to make sounds can be arranged and displayed. As the instruments and items will be used frequently, the display will continually change. Encourage the children to replace items carefully and tidily. In turn, create opportunities for pairs of children to take responsibility for organising the items. Provide labels for the names of the instruments, as well as vocabulary cards that can be used for matching and grouping.

Suggestions for display for each section

Listening for sounds
■ Begin by building up a montage of pictures to represent sound sources the children can identify.
■ Display the observation sheets from Activity 1.
■ Show an illustration of a scene such as a farmyard or busy road and write labels for the sound sources radiating around the picture.
■ Make six boxes or areas to represent *loud, quiet, high, low, all the time, now and again* in which to include pictures and words representing these descriptions of sounds.
■ Display the children's artwork that describes music they have listened to. Provide appropriate captions.

Making sounds
■ Use strings of letters representing sounds as borders surrounding a display of the children's work.
■ Arrange cartoon figures or the children's drawings to demonstrate making sounds with different parts of the body.
■ Have a parade of heads showing different facial expressions made when making sounds such as whistling, humming and singing. Add any photographs of the children in these poses.
■ Display pictures of animals and the names of the distinctive sounds they make.

Enjoying sounds
■ Build up a collection of items used to make sounds and classroom instruments. These should be easily available for observing and exploring, grouping and regrouping.
■ Provide vocabulary labels for children to match to instruments.
■ Add pictures of other instruments not available for classroom use.
■ Have a space to show *Sounds we like* and *Sounds we do not like.*

How we hear
■ Show different animal ears and diagrams showing sound reaching our ears.
■ Display information and photographs relating to the test in Activity 3.
■ Show pictorial examples to demonstrate how our ears warn of dangers. Include road safety examples in a prominent position.
■ Use visual techniques to show the loudness of some sounds.

Investigation
■ Create a bulletin board to emphasise the importance of the science investigation.
■ Show the sequences of the investigation and include appropriate vocabulary that the children can refer to when recording.
■ Display the plan, the measurements and the graph.
■ Add any photographs and children's accounts that are appropriate.
■ Print out the conclusion the children arrived at, as well as a simple explanation describing for visitors the purpose of the test and the children's part in the investigation.

Expressive sounds
■ Work from this section can be used to develop the displays arising from Sections 1 to 3.
■ Provide new vocabulary relating to the collection of instruments.
■ Display pictures of conductors working with choirs and orchestras.

A sound story
■ Use lines of the story or poem as a focal point for a display. This will represent the children's efforts to illustrate a story expressively.
■ Highlight examples of sounds and how these are represented.
■ Include any photographs taken during rehearsals, as well as children's illustrations and written accounts.

Assessment

At the end of the topic on *Sound*:

SCIENCE

SCIENCE
- ■ Can the children recognise a wide range of different sounds?
- ■ Can they identify sources of sounds?
- ■ Can they make observations of sounds by listening carefully?
- ■ Are they aware that there are many ways they can make sounds?
- ■ Do they know ways of making sounds using parts of their bodies?
- ■ Can they compare ways of making sounds?
- ■ Are they able to describe different sounds?
- ■ Can they distinguish between pleasant and unpleasant sounds?
- ■ Have they explored how sounds are made through observing and touching?
- ■ Can they explain that they hear sounds through their ears?
- ■ Are they aware that their ears provide information about the direction from which sounds come?
- ■ Can they describe what they observe when they move further away from a source of sound?
- ■ Do they understand that they use their sense of hearing for a range of purposes, including recognising danger?
- ■ Do they know that loud sounds can be heard over a long distance?
- ■ Have they taken part in a class investigation?
- ■ Can they make suggestions and provide ideas for an investigation?
- ■ Can they make observations, measurements and recordings as part of a class investigation?
- ■ Do they recognise the need to present results, make comparisons and draw conclusions to complete an investigation?
- ■ Have they enjoyed using their experience of sound to illustrate a story or poem with appropriate sound effects?

MUSIC

MUSIC
- ■ Can the children identify a wide range of sound sources?
- ■ Have they made observations of sounds by careful listening?
- ■ Can they recognise different sounds in music?
- ■ Have they explored different ways of making sounds using their voices and parts of their bodies?
- ■ Have they explored ways of making sounds with everyday items and musical instruments?
- ■ Can they identify simple musical instruments?
- ■ Have they played instruments in different ways?
- ■ Can they respond to visual signals when singing and playing?
- ■ Are they able to handle and play instruments with control?
- ■ Have they explored and selected sounds to reflect a song's mood?
- ■ Can they identify sound sources in a story or poem?
- ■ Using their experience, can they suggest and use sounds expressively to illustrate a story or poem?
- ■ Have they had the opportunity to practise, rehearse and take part in a class performance?

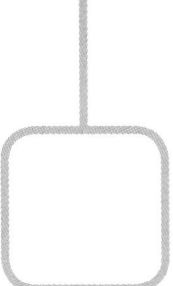

Drawing the topic to a close

With the children, look back over the work they have done and the experiences they have had during the topic on *Sound*. Highlight any special incidents and achievements. Point out how much the children have learned, all the new things they have discovered, the investigations they have made, the games they have played and the enjoyment they have had. Remind them of their new skills with musical instruments; how well they have sung and performed. Talk about the expressive sounds they suggested and practised, culminating in the class performance that is recorded so that others can enjoy it.

Ask the children which parts of the topic they have enjoyed the most and which pieces of work they are most proud of.

Arrange a simple event as a finale. Perhaps the children could perform for another class or for the rest of the school. Invite parents, carers and other family members to visit the class to see the displays of the children's work, along with their individual booklets or folders. Suggest the children act as guides to help any visitors. If appropriate, the children can demonstrate their singing and skills with musical instruments, perhaps performing their sounds to accompany the story or poem.